Contents

3. Attitudes of Preservice Teachers

4. Content and Content/Methods Courses

Preparing Elementary School Mathematics Teachers

Readings from the *Arithmetic Teacher*

edited by

Joan Worth

**Faculty of Education
University of Alberta
Edmonton, Alberta**

National Council of Teachers of Mathematics

Library of Congress Cataloging in Publication Data:

Preparing elementary school mathematics.

   1. Mathematics—Study and teaching (Elementary)
2. Mathematics—Teacher training. I. Worth, Joan.
II. National Council of Teachers of Mathematics.
III. Arithmetic teacher.
QA135.5.P718 1987   372.7   87-34717
ISBN 0-87353-251-1

Printed in the United States of America

5. Topics from Elementary School Mathematics

6. Specific Materials and Techniques

Introduction

Being a teacher educator, like being a teacher, is a rather lonely occupation. Most of our day is spent working in isolation from our peers. Thus, we have limited opportunities to learn from one another. This publication is intended to help us profit from the experience and wisdom of our colleagues. It brings together articles from the *Arithmetic Teacher* on preservice elementary school mathematics teacher education, prepared by and for those who teach intending teachers of elementary school mathematics. Twice in the past the *Arithmetic Teacher* has highlighted articles dealing with teacher education. "Forum on Teacher Preparation" appeared in volumes 15–18, and a series called "Teacher Education" was featured in volume 31. Many of the articles from these volumes are reproduced here.

This collection of articles has two specific purposes. One is to serve as a reminder of the broader issues and concerns that must be considered when preparing elementary school mathematics teachers. The other is to provide a source of ideas and activities for use with classes of preservice teachers.

The first purpose guided the choice of articles for sections one through three, which deal with the teacher educator as teacher, with perspectives on planning an elementary school mathematics teacher preparation program, and with the attitudes of preservice teachers. The remaining three sections focus on how to teach preservice teachers how to teach mathematics to elementary school children. The articles contain many suggestions for helping preservice teachers develop effective mathematical pedagogy. Two themes emerge from these sections. One is that preservice teachers can learn mathematics in a way that is relevant to, and useful in, teaching mathematics to children. The other is that adults as well as children can learn better by doing things than by being told about them. Both these themes are rooted in the assumption that beginning teachers will teach not only what they are taught but as they are taught. Teacher educators, then, must act as role models or exemplars, providing learning experiences that their students can translate into behaviors appropriate for an elementary school teacher.

It is hoped that these readings will be useful not only in teaching undergraduates but also as a resource for graduate students who aspire to become elementary school mathematics teacher educators.

Mathematics Educators as Teachers

THE teacher is the key figure in a child's mathematical education; the teacher educator plays a similar role in a preservice teacher's mathematical education. How teacher educators teach may have as great an impact as what they teach. It is often said that teachers will teach as they were taught. Therefore, teacher educators must model the kinds of teaching behaviors they expect their students to use in their own classrooms.

The articles in this section focus on the mathematics educator as teacher. In "A Prescription for Dedicated Teachers," Frye offers twelve suggestions for becoming an enthusiastic and stimulating teacher that are as applicable to the teacher educator as they are to the elementary school teacher.

Two research reports follow: Suydam describes behaviors of effective teachers in "Teaching Effectiveness"; Leinhardt and Putnam's "Profile of Expertise in Elementary School Mathematics Teaching" reviews research dealing with the contrast between novice and expert teachers of elementary school mathematics.

In the final article, "What Every Elementary School Mathematics Teacher Should Read—Twenty-two Opinions," Leake presents the results of a survey of mathematics educators to discover which ten books on the teaching of elementary school mathematics they would suggest that new teachers purchase for their personal libraries.

One Point of View

A Prescription for Dedicated Teachers

By **Shirley M. Frye**
Scottsdale School District, Scottsdale, AZ 85018

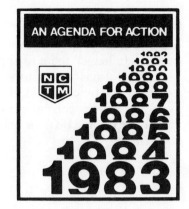

AN AGENDA FOR ACTION

NCTM

1983

The discussion of Recommendation 7 of *An Agenda for Action* (1980) includes this statement: "Even the best prepared, competent, and dedicated teachers must continue their development to keep abreast of changing needs, tools, and conditions." This development could be accomplished by using the "preventive maintenance" theory advocated by home builders, dieticians, automobile mechanics, dentists, and others. Just as these specialists prepare a preventive prescription for their clientele, I have a prescription for our development and growth to keep us excited and exciting in our teaching.

Rx: A Set of Be-attitudes

B_1—Be a teacher.

Be the catalyst for learning as you promote discussion, introduce ideas, clarify concepts, organize the classroom environment, transmit information, and motivate your students.

B_2—Be enthusiastic.

Establish in your classroom a climate for enjoying the challenge of work, and let students see that you enjoy your work. Your enthusiasm is contagious, but so is your indifference!

B_3—Be a positive reinforcer.

Find something reinforcing to say to each of your students every day! Create a positive learning atmosphere in which questioning is acceptable and an "I don't know" prompts assistance. You can relieve math anxiety among your students by preventing negative experiences and by eliminating the fear of our favorite subject.

B_4—Be a change agent.

Alter your room and furniture, change the order of your presentation, vary kinds of materials, and switch the schedule of the day. After you try new approaches, incorporate or modify those that do produce results and discard those that do not. Ideas that don't succeed are not necessarily failures—they just don't fit the situation, the moment, or your style.

B_5—Be creative.

Involve problem solving in each day's lessons in your own creative way. Ask thought-provoking questions, group students for projects, provide unique and stimulating activities, and integrate mathematics into science, social studies, and language arts.

B_6—Be a student.

Be a role model of a constant learner. Let your student see you reading, solving a puzzle, pondering over a mathematical problem, and learning the skills of this technological age. Be the living example of a student for them to follow.

B_7—Be daring.

Put away the pencils, papers, packets, and ditto sheets to allow students the opportunity to talk mathematics. Have students work with their peers and teach each others as they discuss alternatives, procedures, and algorithms. Mathematics is a language that requires dialogue, not just monologue by the teacher.

B_8—Be flexible.

Differentiate the assignments for different groups of students on the basis of their results on your tests. Use evaluation and diagnosis as tools to help tailor the practice, applications, and extension assignments for the individuals.

B_9—Be selective.

Study the multitude of available resources and select those that fit your teaching style. Choose strategies and materials that meet your students' needs. Orchestrate your repertoire of teaching skills to fit your current class, and maintain an organized retrieval system to support you.

B_{10}—Be prepared.

Be ready to teach for the future, which is where your students will need the skills that they learn today. Now is the time to incorporate the calculator, the computer, the video disk, television, and other tools into your programs to enhance your delivery system. Technology will not wait for us to catch up!

B_{11}—Be professional.

Join your local mathematics organization and the NCTM to benefit from professional associations. Read the journals, periodicals, and academic literature so that you are aware of current trends in teaching and research. Consider what recent research reveals about when and how children learn mathematics.

B₁₂—Be a cupid.

Bring your students to a love of mathematics. Influence them to appreciate its contributions to history, to be aware of its functions in their own lives, and to be cognizant of its potential for the future. Be their mathematics mentor!

This is my prescription for the continued development that is required of all professional teachers in this decade of action. I challenge you to try B_1 to B_{12} to maintain your highly effective performance that earns you the reputation of a *dedicated professional teacher!*

Reference

National Council of Teachers of Mathematics. *An Agenda for Action: Recommendations for School Mathematics of the 1980s.* Reston, Va.: The Council, 1980.

The Editorial Panel encourages readers to send their reactions to the author with copies to NCTM (1906 Association Drive, Reston, VA 22091) for consideration in "Readers' Dialogue." ☛

Research Report

Teaching Effectiveness

By **Marilyn N. Suydam**
Ohio State University,
Columbus, OH 43212

Have you ever wondered what teachers do that makes them effective? Suppose effective teachers are those whose students achieve higher test scores. Certain ways in which such teachers manage their classes are different from the ways less effective teachers handle their classes. For instance, effective teachers are much more likely to—

• let pupils know that they are expected to learn and that the teacher is concerned about their achievement;

• give encouragement to pupils, especially about their learning of mathematics;

• ask frequently for pupils' comments and questions;

• minimize waste time, for instance, by planning efficient transitions, giving clear directions, allowing few distractions and interruptions;

• establish and follow simple, consistent rules, taught systematically;

• monitor pupils' behavior carefully and stop inappropriate behavior quickly;

• move around the classroom to monitor pupils' behavior and to give help; and

• give clear directions.

Many other factors are emerging from classroom observations. If you'd like to read more about such ideas, a few sources are given in the Bibliography.

Bibliography

Emmer, Edmund T., Carolyn M. Evertson, and Linda M. Anderson. "Effective Classroom Management at the Beginning of the School Year." *Elementary School Journal* 80 (May 1980):219–31.

Evertson, Carolyn M., and Edmund T. Emmer. "Effective Management at the Beginning of the School Year in Junior High Classes." *Journal of Educational Psychology* 74 (August 1982):485–98.

Rosenshine, Barak. "Teaching Functions in Instructional Programs." *Elementary School Journal* 83 (March 1983):335–51.

Wyne, Marvin D., and Gary B. Stuck. "Time and Learning: Implications for the Classroom Teacher." *Elementary School Journal* 83 (September 1982):67–85. ☛

Research Report

Profile of Expertise in Elementary School Mathematics Teaching

Over the last five years considerable interest and research have been generated about the contrast between novice and expert teachers of elementary school mathematics. Researchers hope that making this contrast will reveal important insights into the nature of expertise—what it means to be an expert mathematics teacher. The research has its roots in a long and distinguished program of psychological studies conducted by Nobel laureate Herbert Simon and others (de Groot 1966; Simon and Chase 1973). The basic approach is to observe closely one or more experts and one or more novices performing the same set of tasks and to look for similarities and differences in the performance of the two groups. Experts and novices usually perform similarly on some of the tasks and quite differently on other, closely related tasks. Contrasting the performance of experts with

Edited by **Leroy G. Callahan**
State University of New York at Buffalo
Buffalo, NY 14260

Prepared by **Gaea Leinhardt** and
  Ralph R. Putnam
University of Pittsburgh
Pittsburgh, PA 15260

that of novices can reveal much of the implicit, or hidden, knowledge of the expert that might otherwise go unnoticed. An example may be helpful. In one study (Simon and Chase 1973), a master chess player (the "expert") and two less competent players (the "novices") looked at a midgame chess board for only a few seconds and then reconstructed the arrangement of the pieces on an empty board. The expert was considerably more successful than the novices. Interestingly, however, when the task was to reconstruct a *random* arrangement of chess pieces, the expert was no better than novices. The expert appeared to have built up a system of knowledge that allowed him to recognize familiar patterns of chess pieces, thus greatly increasing his memory.

In applying this general expert-novice approach to the study of teachers, two serious obstacles have to be overcome: first, defining an expert teacher, and, second, deciding what tasks should serve as the basis for comparing experts and novices. Currently, at least three independent programs of research on teaching are using some form of the expert-novice approach.

The oldest program was started in 1981 at the University of Pittsburgh's Learning Research and Development Center. Expertise in mathematics teaching was defined in this project by the success of the teachers' students on standardized achievement tests.

Teachers were selected as experts if, over a five-year period, their students were among the top classrooms in terms of *growth* on achievement-test performance; they were also confirmed as experts by local supervisors and principals. Their teaching was monitored for one to three years using videotapes and interviews with both children and teachers. The novices were preservice teachers. Analysis focused on the structure of lessons, the nature of mathematical explanations, and the students' grasp of the material taught (Leinhardt 1986b; Leinhardt and Greeno 1986; Leinhardt and Smith 1985; Putnam and Leinhardt 1986).

A second program of research was started at Stanford University by Shulman. In this project, contrasts were made between teachers who are teaching in their own area of expertise and the same teachers teaching in areas in which they are competent but not expert. For example, in one study a biology teacher was monitored as she taught both biology and physics. The results will help us understand how teachers deal with familiar and unfamiliar content (Hashweh 1986; Shulman 1985).

A third research program, initiated by Berliner and Carter at the University of Arizona in 1985, has focused on the basic psychological features of experts, novices, and postulants. Experts were chosen primarily by nomination and observer selection; novices were preservice teachers; and postulants were individuals with training in mathematics or science who were working in other fields but had expressed an interest in becoming teachers. Each group was given a va-

Experts have more elaborate and flexible lessons.

riety of simulation tasks, including planning a lesson given a topic and student records and responding from memory to photos of different classrooms (similar to the chess-board task described earlier). The findings have pointed to many distinctions among these teachers. In general, expert teachers tended to show greater depth in their reports and plans and tended to ignore prior information about students, preferring to form their own opinions (Berliner 1986).

The results of all these studies reveal a profile of expertise in elementary mathematics instruction. Expert teachers have specialized knowledge about the specific topics they teach, about students, and about the nature of cohesive, well-structured lessons. Their knowledge about mathematical topics is specifically focused on the *teaching* of those topics. Expert teachers give more elaborate and flexible lessons (Shulman 1985). They display this knowledge by constructing lessons in which multiple representations of the concept or procedure to be taught are developed. For example, for teaching fractions an expert teacher might use a geometric representation, a number-line representation, and a display of discrete objects. Experts also know different ways of explaining a particular topic, although they are careful not to confuse students by using them all at once (Leinhardt and Smith 1985; Leinhardt and Greeno 1986; Leinhardt 1986a).

The explanations that experts offer are cohesive and tightly connected to the representations being used. An expert using felt strips and craft sticks to explain regrouping will explicitly show that they are being used to demonstrate the same idea. The language of experts' explanations is also cleaner, with the precise use of terms and an avoidance of multiple meanings. Novices' explanations, in contrast, are often perfectly intelligible to someone who already knows the topic but are incomplete and disjointed to a student first learning the material.

Teachers develop mental plans that we call *agendas;* these are not formal written lesson plans but mental notes about what is to happen in the upcoming lesson. Novices have brief, incomplete mental notes with little or no overarching structure; experts' agendas are also brief but intricate and highly structured. They provide a logical flow for the lesson and include references to the actions students are to carry out, to the teacher's actions, and to test points in the lesson (where decisions are made about whether to backtrack or to go on) (Leinhardt 1986b). For topics they teach often, experts have *curriculum scripts* that they continuously refine (Putnam, in press; Putnam and Leinhardt 1986). Expert teachers form lesson plans faster than do novices, and the resulting plans are more flexible and more focused on the content to be taught (Berliner 1986).

In carrying out their agendas, or plans, experts are better at juggling the multiple goals of covering the mathematical content and making sure children are grasping the material. They are able to complete the explanations they offer and maintain consistency within those explana-

tions. Novices, however, often leave their explanations incomplete or collide with the clock and end up having students "finish" the lesson as homework, something experts rarely do (Leinhardt and Greeno 1986).

The goal of analyzing the teaching performance of experts is to find ways to help all teachers, especially novices, improve their teaching—to become experts themselves. To become successful, the novice teacher of mathematics must build up a specialized kind of mathematical knowledge—knowledge that is specifically tied to

---

Experts use terms in a precise manner.

---

teaching (Leinhardt and Greeno 1986; Shulman 1986). This base of mathematics knowledge *for instruction* provides the multiple representations and explanations of various topics that expert teachers use. This research holds promise for offering many more specific recommendations and perspectives useful to teachers trying to improve their professional skills.

References

Berliner, David. "In Pursuit of the Expert Pedagogue." Presidential address to American Educational Research Association, San Francisco, Calif., April 1986.

de Groot, Adriaan D. "Perception and Memory versus Thought: Some Old Ideas and Recent Findings." In *Problem Solving: Research, Method and Theory,* edited by B. Kleinmuntz. New York: John Wiley & Sons, 1966.

Hashweh, Maher Z. "Effects of Subject-Matter Knowledge in the Teaching of Biology and Physics." Paper presented at the American Educational Research Association meeting, San Francisco, Calif., April 1986.

Leinhardt, Gaea. "Expertise in Mathematics Teaching." *Educational Leadership* 43 (April 1986a):28–33.

———. "Math Lessons: A Contrast of Novice and Expert Competence." Paper presented at the American Educational Research Association meeting, San Francisco, Calif., April 1986b.

Leinhardt, Gaea, and James Greeno. "The Cognitive Skill of Teaching." *Journal of Educational Psychology* 78 (April 1986):75–95.

Leinhardt, Gaea, and Donald Smith. "Expertise in Mathematics Instruction: Subject Matter Knowledge." *Journal of Educational Psychology* 77 (June 1985):247–71.

Putnam, Ralph T. "Structuring and Adjusting Content for Students: A Study of Live and Simulated Tutoring of Addition." *American Educational Research Journal*, in press.

Putnam, Ralph T., and Gaea Leinhardt. "Curriculum Scripts and the Adjustment of Content in Mathematics Lessons." Paper presented at the American Educational Research Association meeting, San Francisco, Calif., April 1986.

Shulman, Lee S. "Knowledge Growth in Teaching: New Approaches to an Old Problem." Symposium conducted at the American Educational Research Association meeting, Chicago, Ill., April 1985.

———. "Those Who Understand: Knowledge Growth in Teaching." *Educational Researcher* 15 (February 1986):4–14.

Simon, Herbert A., and William G. Chase. "Skill in Chess." *American Scientist* 61 (July/August 1973):394–403. ✏

What Every Elementary School Mathematics Teacher Should Read— Twenty-two Opinions

By **Lowell Leake**

If you could own only ten books on the teaching of elementary school mathematics, which would you choose? If your school's library suddenly acquired enough money to buy ten such books, which would you recommend? Which ten books on the subject would you suggest that new teachers purchase for their personal libraries? Such questions are more limited versions of the old idea of deciding which ten books you would want if you were marooned on a desert island.

This article reports the results of a survey that asked twenty-nine distinguished mathematics educators to list the ten most important or useful books for a classroom teacher of elementary school mathematics (K–6) to read. Three similar surveys have been reported in the *Mathematics Teacher* (Lazarnick and Frantz 1984; Leake 1972; and Leake 1983), but this was the first to focus on the elementary school classroom.

Of the twenty-nine educators surveyed, seventeen are women and twelve are men. Twenty-two replied, thirteen women and nine men. They are Gary G. Bitter, Elizabeth Fennema, David Fielker, Shirley Frye, Carole Greenes, Judith Jacobs, Martin Johnson, Margaret J. Kenney, Betty K. Lichtenberg, Mary M. Lindquist, Bonnie H. Litwiller, Lola J. May, Doyal Nelson, Phares G.

Lowell Leake teaches at the University of Cincinnati, Cincinnati, OH 45221. His interests and activities focus on the improvement of training for mathematics teachers at all levels and on the role of the computer in this mathematics education.

O'Daffer, Teri H. Perl, Len Pikaart, G. Edith Robinson, Thomas A. Romberg, William E. Schall, Hilary Shuard, Marilyn N. Suydam, and J. Fred Weaver. Two prominent mathematics educators from England were included (David Fielker and Hilary Shuard) to give the survey some input from abroad. All the other respondents are from the United States and Canada.

The books named are listed in eleven categories; they are arranged alphabetically within each category. Many of the books, of course, could easily have appeared in several categories. Following the bibliographical information for each book is a list of the respondents who chose that book. The inclusion of a price means the book is in print.

A few general observations can be made about the list. First, the variety of books named is extremely wide; 115 books are listed, which is roughly comparable to the number of books named in the earlier surveys. The most notable difference from my earlier surveys (Leake 1972, 1983) is the appearance of a number of titles in the category "Women and Mathematics." Six titles were listed in this category nine times by five educators. Otherwise, the categories are similar to those of the previous surveys. Readers will no doubt observe some other generalizations that I have omitted or overlooked.

What were the most popular books? The eleven titles that follow were listed five times or more:

- *Mindstorms: Children, Computers, and Powerful Ideas,* Seymour Pa-

pert: twelve times
- *Problem Solving in School Mathematics,* 1980 Yearbook of the NCTM: nine times
- *Freedom to Learn: An Active Learning Approach to Mathematics,* Edith E. Biggs and James R. MacLean: eight times
- *Developing Computational Skills,* 1978 Yearbook of the NCTM: seven times
- *An Agenda for Action: Recommendations for School Mathematics of the 1980s,* NCTM: six times
- *Children's Arithmetic: The Learning Process,* Herbert Ginsburg: six times
- *Mathematics Learning in Early Childhood,* Thirty-seventh Yearbook of the NCTM: six times
- *Research within Reach: Elementary School Mathematics,* Mark J. Driscoll: six times
- *Selected Issues in Mathematics Education,* Mary M. Lindquist: six times
- *Mathematics, a Human Endeavor: A Book for Those Who Think They Don't Like the Subject,* Harold R. Jacobs: six times
- *Mathematics Their Way,* Mary Baratta-Lorton: five times

Finally, the NCTM can take pride in the fact that seventeen of its publications were named fifty-eight times.

Computers and Calculators

Activities Handbook for Teaching with the Hand-held Calculator, Gary G. Bitter and Jerald L. Mikesell. Boston: Allyn & Bacon, 1979. 336 pp., $18.95 [Bitter]

An Apple for the Teacher: Fundamentals of Instructional Computing, George C. Culp and Herbert Nickles. Belmont, Calif.: Brooks/Cole Publishing Co., 1983. 239 pp., $15.95. [Litwiller]

Arithmetic and Calculators: How to Deal with Arithmetic in the Calculator Age, William G. Chinn, Richard A. Dean, and Theodore N. Tracewell. San Francisco: W. H. Freeman & Co., 1978. 488 pp., $25.25. [Kenney]

BASIC: A Hands-On Method, 2d ed., Herbert C. Peckham. New York: McGraw-Hill, 1981. 320 pp., $12.95. [Schall]

Calculators in the Primary Schools, Milton Keynes. Nottingham, England: Open University, 1982. 83 pp. [Shuard]

Computers in Today's World, Gary G. Bitter. New York: John Wiley & Sons, 1983. [Bitter]

Mindstorms: Children, Computers, and Powerful Ideas, Seymour Papert. New York: Basic Books, 1980. 230 pp. $14.95. [Fielker, Greenes, Jacobs, Johnson, Kenney, Lindquist, May, O'Daffer, Perl, Robinson, Schall, Shuard]

Spotlight on Computer Literacy, Ellen Richman. New York: Random House, 1982. 186 pp. [Lichtenberg]

Using a Microcomputer in the Classroom, Gary G. Bitter and Ruth A. Camuse. Reston, Va.: Reston Publishing Co., 1984. 339 pp. [Bitter]

Content

A Problem Solving Approach to Mathematics for Elementary School Teachers, Rick Billstein, Shlomo Libeskind, and Johnny W. Lott. Menlo Park, Calif.: Benjamin/Cummings Publishing Co., 1981. 656 pp., $21.95. [Kenney]

Experiencing Geometry, James V. Bruni. Belmont, Calif.: Wadsworth Publishing Co., 1977. 310 pp., $15.95. [Frye, Robinson]

From Sticks and Stones: Personal Adventures in Mathematics, Paul B. Johnson, Chicago: Science Research Associates, 1975. 552 pp., $17.95. [Schall]

Geometry: An Investigative Approach, Phares O'Daffer and Stanley R. Clemens. Menlo Park, Calif.: Addison-Wesley Publishing Co., 1976. 445 pp. [Greenes, Kenney]

Mathematics, a Human Endeavor: A Book for Those Who Think They Don't Like the Subject, Harold R. Jacobs. San Francisco: W. H. Freeman & Co., 1982. 649 pp., $17.95. [Kenney, Lichtenberg, O'Daffer, Perl, Pikaart, Schall]

Mathematics and the Imagination, Edward Kasner and J. R. Newman. New York: Simon & Schuster, 1962. 380 pp., $4.95 [Perl]

Mathematics and the Physical World, Morris Kline. New York: Thomas Y. Crowell Co., 1963. 546 pp., $6.00. [Frye]

Mathematics: The Alphabet of Science, Margaret F. Willerding and Harold S. Engelsohn. New York: John Wiley & Sons, 1977. 651 pp. [Lichtenberg]

The Nature and Growth of Modern Mathematics, Edna E. Kramer. Princeton, N.J.: Princeton University Press, 1982. 758 pp., $45.00. [Perl]

Patterns and Systems of Elementary Math, Jonathan Knaupp. Boston: Houghton Mifflin Co., 1977. 425 pp. [Bitter]

Vision in Elementary Mathematics, W. W. Sawyer. Baltimore: Penguin Books, 1964. 346 pp. [Perl]

What Is Mathematics?, Richard Courant and Herbert Robbins. Fairlawn, N.J.: Oxford University Press, 1941. 521 pp., $8.00 [Robinson]

Curriculum Surveys and Recommendations

An Agenda for Action: Recommendations for School Mathematics of the 1980s. Reston, Va.: NCTM, 1980. 29 pp., $1.00. [Jacobs, Lichtenberg, Lindquist, O'Daffer, Shuard, Suydam]

The Agenda in Action, 1983 Yearbook of the NCTM, Gwen Shufelt, ed. Reston, Va.: The Council, 1983. 245 pp., $14.50. [Litwiller, Schall]

Education in the 80's: Mathematics, Shirley Hill, ed. Washington, D.C.: National Education Association, 1982. 120 pp. [Litwiller]

Mathematical Development: Primary Survey Report No. 3, Assessment of Performance Unit. London: H.M.S.O., 1980. 162 pp., $37.00. [Shuard]

Mathematics Counts, Report of the Committee of Inquiry into the Teaching of Mathematics in Schools under the Chairmanship of W. H. Crockroft. London: H.M.S.O., 1982. 311 pp. [Fielker, Shuard]

Mathematics 5–11, Her Majesty's Inspectors of Schools. London: H.M.S.O., 1979. [Shuard]

Overview and Analysis of School Mathematics, K–12, Shirley Hill, chr. Washington, D.C.: Conference Board of Mathematical Sciences, National Advisory Committee on Mathematics Education, 1975. Available from NCTM. 157 pp., $3.40. [Johnson]

Exceptional Children

The Intellectually Gifted: An Overview, James J. Gallagher. New York: Grune & Stratton, 1976. [Greenes]

Mathematical Education of Exceptional Children and Youth: An Interdisciplinary Approach, Vincent J. Glennon, ed. Professional Reference Series. Reston, Va.: NCTM, 1981. 408 pp., $28.00. [Greenes, Johnson, Litwiller]

Teaching Mathematics to Children with Special Needs, Fredericka K. Reisman and Samuel H. Kauffman. Columbus, Ohio: Charles E. Merrill Publishing Co., 1980. 336 pp., $20.50. [Weaver]

History and Philosophy of Mathematics

Godel, Escher, Bach: An Eternal Golden Braid, Douglas R. Hofstadter. New York: Random House, 1979. 772 pp., $9.95. [Pikaart]

History of Mathematics Education in the U.S. and Canada, Thirty-second Yearbook of the NCTM, Phillip S. Jones, ed. Reston, Va.: The Council, 1970. 557 pp., $16.90. [Lindquist, Romberg]

Human Values and Science, Art, and Mathematics, Lillian R. Lieber. New York: W. W. Norton & Co., 1961. 149 pp., $3.95. [Lichtenberg]

An Introduction to the History of Mathematics, 4th ed., Howard Eves. New York: Holt, Rinehart & Winston, 1976. 464 pp., $26.95. [Frye]

Mathematical Thought from Ancient to Modern Times, Morris Kline. New York: Oxford University Press, 1979. 1238 pp., $60.00. [Pikaart]

Mathematics: An Introduction to Its Spirit and Use: Readings from Scientific American, introduction by Morris Kline. San Francisco: W. H. Freeman & Co., 1979. 249 pp., $9.95. [Romberg]

Men of Mathematics, Eric T. Bell. New York: Simon & Schuster, 1937. 592 pp., $12.95. [Frye]

A Survey of Mathematics: Educational Concepts and Their Historical Development, Vivian S. Groza. New York: Holt, Rinehart & Winston, 1968. 327 pp. [Litwiller]

The Whole Craft of Number, Douglas M. Campbell. Boston: Prindle, Weber & Schmidt, 1976. 514 pp., $22.30. [Kenney]

The World of Mathematics, 4 vols., James R. Newmann, ed. New York: Simon & Schuster, 1965. 2525 pp. [Pikaart]

Methodology

Another, Another, Another and More, Marion Walter. London: Andre Deutsch, 1975. 35 pp., $8.75. [Fielker]

Developing Computational Skills, 1978 Yearbook of the NCTM, Marilyn N. Suydam, ed. Reston, Va.: The Council, 1978. 245 pp., $14.50. [Lindquist, Litwiller, May, Nelson, O'Daffer, Suydam, Weaver]

Diagnosing Mathematical Difficulties, Richard G. Underhill, A. Edward Uprichard, and James W. Heddens, eds. Columbus, Ohio: Charles E. Merrill Publishing Co., 1980. 408 pp., $18.95. [Robinson]

Didactics and Mathematics, Mathematics Resource Project, University of Oregon. Palo Alto, Calif.: Creative Publications, 1978. 188 pp., $15.50. [Schall, Suydam]

Elementary Arithmetic, Its Meaning and Practice, Burdett R. Buckingham. Boston: Ginn & Co., 1947. 744 pp. [Nelson]

Elementary Mathematics Today: A Resource for Teachers, Elizabeth Williams and Hilary Shuard. Menlo Park, Calif.: Addison-Wesley Publishing Co., 1976. 455 pp. [Frye, Greenes, Perl, Romberg]

Enrichment Mathematics for the Grades, Twenty-seventh Yearbook of the NCTM. Washington, D.C.: The Council, 1963. 368 pp., $14.60. [May]

Error Patterns in Computation, 3d ed., Robert B. Ashlock. Columbus, Ohio: Charles E. Merrill Publishing Co., 1982. 208 pp., $8.95. [Greenes, Jacobs, Weaver]

Freedom to Learn: An Active Learning Approach to Mathematics, Edith E. Biggs and James R. MacLean. Reading, Mass.: Addison-Wesley Publishing Co., 1969. 206 pp., $14.00. [Bitter, Fielker, Frye, Johnson, May, O'Daffer, Robinson, Schall]

Geometry in the Classroom, Harold A. Elliot, James R. MacLean, and Janet M. Jordan. Toronto: Holt, Rinehart & Winston of Canada, 1968. 266 pp., $5.01. [O'Daffer]

The Growth of Mathematical Ideas Grades K–

12, Twenty-fourth Yearbook of the NCTM, Phillip S. Jones, ed. Washington, D.C.: The Council, 1959. 507 pp., $7.00. [Frye, Pikaart]

A Guide to the Diagnostic Teaching of Arithmetic, 3d ed., Fredericka Reisman. Columbus, Ohio: Charles E. Merrill Publishing Co., 1982. 184 pp., $8.95. [Greenes]

Guiding Chidren to Mathematical Discovery, 3d ed., Leonard M. Kennedy. Belmont, Calif.: Wadsworth Publishing Co., 1980. 533 pp., $24.95. [Nelson]

Guiding Each Child's Learning of Mathematics: A Diagnostic Approach to Instruction, R. Ashlock, M. Johnson, J. Wilson, and W. Jones. Columbus, Ohio: Charles E. Merrill Publishing Co., 1983. 507 pp. [Johnson]

Helping Children Learn Mathematics, Robert Reys, Marilyn N. Suydam, and Mary M. Lindquist. Englewood Cliffs, N.J.: Prentice-Hall, 1984. 384 pp., $24.95. [Lindquist]

Inside the Primary Classroom, Maurice Galton, Brian Suman, and Paul Croll. Boston: Routledge & Kegan Paul, 1980. 203 pp., $25.00 [Shuard]

Lab Manual for Elementary Mathematics, 2d ed., William M. Fitzgerald. Boston: Prindle, Weber & Schmidt, 1973. 178 pp. [Bitter]

Math Activities for Child Involvement, 3d ed., C. W. Schminke and Enoch Duman. Boston: Allyn & Bacon, 1981. 336 pp., $20.95. [Lichtenberg]

Mathematical Experiences for Young Children, Louise B. Scott and Jewell Garner. St. Louis: McGraw-Hill, 1978. 311 pp. [Bitter]

Mathematics, a Good Beginning: Strategies for Teaching Children, 2d ed., Andria P. Troutman and Betty K. Lichtenberg. Monterey, Calif.: Brooks/Cole Publishing Co., 1982. 502 pp., $18.95 [Schall]

Mathematics for the Middle Grades (5–9), 1982 Yearbook of the NCTM, Linda Silvey, ed. Reston, Va.: The Council, 1982. 246 pp., $14.50 [Lichtenberg, Pikaart, Robinson, Suydam]

Mathematics Learning in Early Childhood, Thirty-seventh Yearbook of the NCTM, Joseph N. Payne, ed., Reston, Va.: The Council, 1975. 316 pp., $16.90. [Lichtenberg, Lindquist, Nelson, Pikaart, Robinson, Suydam]

Mathematics Resource Project, A. Hoffer, ed. Palo Alto, Calif.: Creative Publications, 1978.
1. *Number Sense and Arithmetic Skills*, 832 pp., $32.50.
2. *Ratio, Proportion and Scaling*, 516 pp., $25.00.
3. *Geometry and Visualization*, 830 pp., $32.50.
4. *Mathematics in Sciences and Society*, 464 pp., $25.00.
5. Statistics and Information Organization, 850 pp., $32.50. [Pikaart]

Mathematics Their Way, Mary Baratta-Lorton. Menlo Park, Calif.: Addison-Wesley Publishing Co., 1976. 396 pp., $25.80. [Greenes, Jacobs, May, O'Daffer, Suydam]

Mathmatters, Randall J. Souviney, Tamara Keyser, and Alan Sarver. Glenview, Ill.: Scott Foresman & Co., 1978. $13.95. [Bitter]

The Mathworks: Handbook of Activities for Helping Students Learn Mathematics, Carole Greenes, Frank Cristiano, Justin Dombrowski, Eileen Geller, Walter Hayes, Linda Schulman, Eileen Singer, Rita Spungin, Paula Wolf, and Maureen Zolubof. Palo Alto,

Calif.: Creative Publications, 1979. 440 pp., $19.95. [Greenes]

Measurement in School Mathematics, 1976 Yearbook of the NCTM, Doyal Nelson, ed. Reston, Va.: The Council, 1976. 244 pp., $13.75. [Bitter]

Notes on Mathematics for Children, Association of Teachers of Mathematics, D. H. Wheeler, ed. Cambridge: Cambridge University Press, 1977. $28.95, $10.95 paper. [Fielker, Robinson]

Notes on Mathematics in Primary School, Association of Teachers of Mathematics. Cambridge: Cambridge University Press, 1969. 340 pp. [Fielker, Nelson]

Teaching Mathematics in the Elementary Schools, David J. Fuys and Rosamund R. Tischler. Boston: Little, Brown & Co., 1979. 593 pp. [Jacobs, Weaver]

Teaching Statistics and Probability, 1981 Yearbook of the NCTM, Albert P. Shulte, ed. Reston, Va.: The Council, 1981. 246 pp., $13.75. [Kenney]

Thinking Is Child's Play, Evelyn Sharpe. New York: Avon Books, 1970. 143 pp., $1.95 [Schall]

Workjobs: Activity-centered Learning for Early Childhood Education, Mary Baratta-Lorton. Menlo Park, Calif.: Addison-Wesley Publishing Co., 1972. 255 pp., $12.50. [Perl]

Problem Solving

The Book of Think: How to Solve a Problem Twice Your Size, Marilyn Burns. Boston: Little, Brown & Co., 1976. 125 pp., $8.95. [Lichtenberg]

How to Solve It, George Polya. Princeton, N.J.: Princeton University Press, 1971. 253 pp., $4.95. [Frye, Litwiller]

Problem Solving: A Basic Mathematics Goal, Steven Meiring. Columbus, Ohio: Ohio Department of Education, 1980. [Kenney, Pikaart, Suydam]
1. *Becoming a Better Problem Solver*, 61 pp.
2. *A Resource for Problem Solving*, 88 pp.

Problem Solving: A Handbook for the Teacher, Stephen Krulik and Jesse A. Rudnick. Boston: Allyn & Bacon, 1980. 227 pp., $18.80. [Bitter, May]

Problem Solving in School Mathematics, 1980 Yearbook of the NCTM. Stephen Krulik, ed. Reston, Va.: The Council, 1980. 241 pp., $14.50. [Jacobs, Kenney, Lindquist, Litwiller, O'Daffer, Robinson, Schall, Suydam, Weaver]

Psychology of Learning Mathematics

The Child's Conception of Number, Jean Piaget. New York: W. W. Norton & Co., 1965. 248 pp., $4.25. [Johnson, Nelson]

The Child's Understanding of Number, Rochel Gelman and C. R. Gallistel. Cambridge: Harvard University Press, 1978. 260 pp., $16.50. [Nelson]

Children's Arithmetic: The Learning Process, Herbert Ginsburg. New York: D. Van Nostrand Co., 1977. 197 pp. [Fennema, Greenes, Johnson, Romberg, Shuard, Weaver]

Children's Minds, Margaret Donaldson. New York: W. W. Norton & Co., 1979. 166 pp., $3.95. [Fielker, Weaver]

Children's Understanding of Mathematics 11–16, CSMS Mathematics Team, K. M. Hart, et al. London: John Murray, 1981. 231 pp. [Shuard]

Evaluating the Quality of Learning, John B. Biggs and Kevin F. Collis. New York: Academic Press, 1982. 245 pp., $26.50. [Romberg]

Experiences in Visual Thinking, Robert H. McKim. Belmont, Calif.: Brooks/Cole Publishing Co., 1980. 183 pp., $15.95. [Perl]

How Children Fail, John C. Holt. New York: Dell Publishing Co., 1964. 298 pp., $1.95 [O'Daffer]

How Children Learn Mathematics: Teaching Implications of Piaget's Research, Richard W. Copeland. New York: Macmillan, 1978. 449 pp., $19.95. [Frye]

Mathematics and Children, Madeline Goutard. Reading, England: Education Explorers, 1968. 196 pp. [Fielker]

Mathematics Learning in Early Childhood, Thirty-seventh Yearbook of the NCTM, Joseph N. Payne, ed. Reston, Va.: The Council, 1975. 316 pp., $16.90. [Frye, Lichtenberg, Nelson, Pikaart, Robinson, Suydam]

Perception and Understanding in Young Children: An Experimental Approach, Peter E. Bryant. New York: Basic Books, 1974. 195 pp., $7.95. [Nelson]

Piaget for the Classroom Teacher, Barry J. Wadsworth. New York: Longman, 1979. 303 pp., $11.95. [Bitter]

The Psychology of Learning Mathematics, Richard R. Skemp. Baltimore: Penguin Books, 1971. 319 pp., $2.25. [Fielker, Nelson, Romberg]

The Psychology of Mathematical Abilities in Schoolchildren, V. A. Krutetskii. Chicago: University of Chicago Press, 1978. 417 pp., $6.00. [Johnson]

Research

Active Mathematics Teaching, Thomas L. Good, Douglas A. Grouws, and Howard Ebmeier. New York: Longman, 1983. 246 pp. [Fennema, Romberg]

Activity-based Learning in Elementary School Mathematics: Recommendations from Research, Marilyn N. Suydam and Jon L. Higgins. Columbus, Ohio: ERIC, 1977. 178 pp. [Fennema]

Critical Variables in Mathematics Education, E. G. Begle. NCTM and Mathematical Association of America, 1979. 165 pp., $8.00. [Pikaart]

Research within Reach: Elementary School Mathematics, Mark J. Driscoll. Reston, Va.: NCTM, 1981. 141 pp., $6.25. [Fennema, Lichtenberg, Lindquist, Romberg, Suydam, Weaver]

Mathematics Education Research: Implications for the 80's, Elizabeth Fennema, ed. Reston, Va.: NCTM, 1981. 182 pp., $6.75. [Fennema, Lindquist, Litwiller, Romberg]

Research in Mathematics Education, Richard J. Shumway, ed. Reston, Va.: NCTM, 1980. 480 pp., $27.00. [Nelson]

Selected Issues in Mathematics Educations, National Society for the Study of Education, Series on Contemporary Education Issues, Mary M. Lindquist, ed. Berkeley, Calif.:

McCutchan Publishing Corp., 1981. 250 pp., $17.50. [Fennema, Lindquist, Litwiller, O'Daffer, Robinson, Weaver]

Teachers Make a Difference, Thomas L. Good, Bruce J. Biddle, and Jere E. Brophy. New York: Holt, Rinehart & Winston, 1975. 271 pp., $5.95. [Suydam]

Using Research: A Key to Elementary School Mathematics, Marilyn N. Suydam and J. Fred Weaver. Columbus, Ohio: ERIC, 1981. [Weaver]

Women and Mathematics

How to Encourage Girls in Math and Science, Joan Skolnick, Carol Langbort, and Lucille Day. Englewood Cliffs, N.J.: Prentice-Hall, 1982. 192 pp., $15.95, $7.95 paper. [Jacobs]

Math Equals: Biography of Women Mathematicians and Related Activities, Teri Perl. Menlo Park, Calif.: Addison-Wesley Publishing Co., 1978. 445 pp., $11.75. [Johnson, Kenney, Perl]

Math for Girls and Other Problem Solvers, Diane Downie, Twila Slesnick, and Jean K. Stenmark. Berkeley, Calif.: Lawrence Hall of Science, 1981. 108 pp. [Jacobs]

Overcoming Math Anxiety, Sheila Tobias. Boston: Houghton Mifflin Co., 1980. 288 pp., $5.95. [Jacobs, Johnson]

Women and the Mathematical Mystique, Lynn H. Fox, Linda Brady, and Dianne Tobin, eds. Baltimore: Johns Hopkins University Press, 1980. 224 pp. [Fennema]

Women, Numbers and Dreams: Biographical Sketches and Math Activities, Teri H. Perl and Joan M. Manning. Menlo Park, Calif.: Learning Co., 1984. 250 pp., $17.00 [Fennema]

Miscellaneous

Helping Children Read Mathematics, Robert B. Kane, Mary Ann Byrne, and Mary Ann Hater. New York: American Book Co., 1974. 150 pp. [Robinson]

Language Teaching and Learning: Mathematics, Mike Torbe, ed. London: Ward Lock Educational, 1982. [Shuard]

Mathematics Illustrated Dictionary: Facts, Figures and People, Including New Math, Jeanne Bendick and Marcia Levin. New York: McGraw-Hill, 1972. $8.95. [Perl]

Mister God, This Is Anna, Fynn. New York: Ballantine Books, 1976. 192 pp., $2.25 paper. [Fielker]

S.P.A.C.E.S. (Solving Problems of Access to Careers in Engineering and Science), Sherry Fraser, project director. Berkeley, Calif.: Lawrence Hall of Science, 1982. [Jacobs]

Search for Solutions, Horace F. Judson. New York: Holt, Rinehart & Winston, 1980. 224 pp., $16.95. [Romberg]

References

Lazarnick, Sylvia, and Marny Frantz. "Female Mathematics Educators Recommend Books." *Mathematics Teacher* 77 (March 1984):233–34.

Leake, Lowell, Jr. "What Every Mathematics Teacher Ought to Read (Seventeen Opinions)." *Mathematics Teacher* 65 (November 1972):637–41.

———."What Every Secondary School Mathematics Teacher Should Read—Twenty-four Opinions." *Mathematics Teacher* 76 (February 1983):128–33. ◆

Preservice Elementary School Mathematics Programs

2

THE scope and sequence of elementary school mathematics teacher education programs have long been of concern. The articles in this section offer some perspectives on what, when, and how the content, methods, and practicum components may be defined and organized.

In "The Sorry State of Mathematics Teacher Education," Rising provides strong arguments for his belief that in 1969 education programs for mathematics teachers were at best inadequate. Many of the problems he noted are still with us eighteen years later, and many of his suggestions for dealing with the problems are still viable. Dossey's "The Current Status of Preservice Elementary Teacher-Education Programs" reports a 1977 study designed to describe the typical preservice elementary school mathematics education program and to establish a baseline for future teacher-education studies. Several discrepancies between what is and what ought to be are noted. Some of these discrepancies are addressed in "Teacher Preparation—a Coordinated Approach" by Burger, Jenkins, Moore, Musser, and Smith, who describe what they consider to be an exemplary program implemented at their university in 1981. In "Preservice Elementary Mathematics Education: A Complete Program," Dossey returns to provide a model for a comprehensive learning experience for K–6 teachers that includes three interacting dimensions—content, methods, and clinical experiences. Berman and Friederwitzer stress balance and complementarity in content and method when examining "The Teacher Educator's Dilemma: Is More Better?"

The remaining two articles deal with the preparation of teachers for particular grade levels and for particular pupils. Isenberg and Altizer-Tuning in "The Mathematics Education of Primary-Grade Teachers" address the need to integrate general professional preparation, mathematics content, and mathematics methods and make suggestions relating to each. The argument that special-education preservice teachers require the same mathematics education program as regular elementary school preservice teachers is advanced by Hofeldt and Hofeldt in "Mathematics for Special Education Teacher Trainees."

The sorry state of mathematics teacher education[1]

GERALD R. RISING

*Gerald Rising is professor of mathematics education
at the State University of New York at Buffalo.
He works extensively with preservice and in-service teachers of
elementary school mathematics.*

The concern I express in this polemic is a simple but important one: *Education programs for mathematics teachers are not only below any reasonably acceptable standard, but they are getting steadily worse!*

As a complex society makes ever increasing demands upon the scientific competence of its citizens, colleges are providing teachers of mathematics with an inadequate preparation to do their job. While mathematicians and classroom teachers are working together to provide strong mathematics texts for students at all levels, the teachers who will be expected to implement those texts are nurtured on programs that are at best oblique to the tasks they face in the classroom.

What is happening?

First, campus schools are being disbanded. Fewer and fewer colleges provide opportunity for students to participate in a teaching environment that is both within the college and at the same time responsible to members of the college staff. This development has a profound effect upon every phase of the preservice teachers' educational program, because without the availability of campus schools, college instructors tend to become even further isolated from the practical world of the classroom teacher. The greatest loss, however, is that of the "teaching laboratory." In a campus school, teachers-to-be have constantly available to them opportunities to relate theory with practice, to interact immediately with experienced classroom teachers, and to participate in the school on a scale appropriate to their progress in the training program. When a campus school is near at hand, it is possible for college teachers to assign specific observations, to arrange for a variety of experiences preliminary to the student's taking full responsibility for a class, such as tutoring individual children, and even to have college students observe one another teaching. Without a campus school, the difficulty of carrying out any of these activities makes them impractical.

Second, classwork associated with mathematical pedagogy (methods) is being reduced, replaced, or misdirected; and the quality of instruction in these courses is

[1] Before writing, I discussed the matter of this article with students in my graduate seminars, and with experienced mathematics teachers in local public, private, and parochial schools and colleges. To them should go credit for many suggestions, anecdotes, even some research, and a critical reading of the first draft. But with me rests full responsibility for the underlying anger at a situation that I feel is now essentially out of control.

declining. Students today have mathematics backgrounds that are much stronger than was the case for the average undergraduate a decade ago. Yet today's students are being given little professional assistance to develop techniques for translating their mathematical knowledge into viable classroom procedures. One result of this is that mathematics is often taught in the elementary school exactly as it was taught to the teacher in the college classroom. An observer who watches the transition from a reading lesson to a mathematics lesson in an elementary classroom cannot fail to notice the difference in instruction. In the former, thoughtful pedagogy—learned in the teacher education program—is evident; in the latter, spoon feeding, lecturing, and other types of inappropriate pedagogy are the rule. It is difficult for the observer to realize that the teacher is the same person in both instances.

Third, the most important single part of teacher education, the supervised teaching experience, has been weakened to the point where it no longer brings into focus (or, perhaps more accurately, reclaims) the total education program. Supervised teaching is being curtailed, often cut to half-day sessions, and in many schools students undertake it while carrying a heavy burden of coursework. Meanwhile the quality of supervision of the student teacher, by both his master teacher in the school and his university supervisor, is severely reduced. Assigned to poor teachers and seldom visited by the college instructor, student teachers gain only a fraction of what they would in a stronger program.

Why is this happening?

While there are a number of different kinds of answers to this question, most of them stem from a single source: the failure of college faculties to respond to the needs of society *when those needs do not coincide with faculty interests.*

As the game is played, teacher education, like undergraduate instruction and

guidance, as well as teaching (as opposed to research), is not in the self-interest of the college professor. Consequently he turns to roles that will "do him some good." The phenomenon is not peculiar to teacher education; in fact it is precisely this irresponsiveness that is at the base of much current student criticism of institutions of higher learning. The cartoon caption: "It is much more fun to be a counselor than a camper," is incisive in this respect.

In the classical sense, the college is a place where students are brought into contact with a community of scholars—their professors. While the "common denominator" influence of mass education mediates against this definition, it is retained in the rationale behind staff evaluation in terms of quality of research. The matter is further complicated by continued specialization, especially true in education, but true as well in substantive departments. In mathematics, for example, the algebraist doesn't speak the language of the topologist. The individual college faculty member often has more in common professionally with people in other states than he has with the man sharing his office. Partly because of staff evaluation on the basis of this, quality is translated to evaluation by quantity, often with macabre results: two books, five articles, and a dissertation is a better record than one book and six articles. In many cases this is but an uncertain measure using fourteen pieces of unread literature and one read by a committee of three.

In recognition of the shortcomings of the quantity standard, colleges turn to a dangerous subjective standard: stature in the college and in the field—often better described as political status and notoriety. Moreover, the lowest of the low on the status scale today is the teacher-education phase of professional education. Thus, that which is highest in importance to teacher apprentices is translated to lowest in political importance in the college scheme by this perversity of professional self-interest.

Consider the typical makeup of a college or department of education. Generally, less than a quarter of the staff is concerned with instruction in all fields and at any level. Other staff members focus their attention upon such peripheral areas as foundations, higher education, school administration, guidance, educational sociology, educational psychology, and curriculum development. Most of these are familiar to the reader, but a few deserve special comment. Educational sociology, for example, and educational psychology are among the lower-caste breeds, with the latter tending to seek stature by deserting education as much as possible and turning to psychology and statistics. Specialists in curriculum development seriously believe that knowledge of a particular subject is not prerequisite to the development of teaching materials for that field. These generalists hold, almost as an article of faith, that curriculum planning in the schools is the responsibility of the classroom teacher. (Pure hokum: the already enervated classroom teacher should concentrate his remaining energies on improvement of instruction.)

So it is that campus schools are disbanded because they draw money, manpower, and energy away from the personal enterprise of a staff majority. Similarly low on the list of rewarded activities is supervision of student teachers. It is not surprising, then, that faculty members turn their energies to other things and transfer supervisory duties to overworked and underpaid graduate assistants whose first concerns are always with their own studies. Reduced interest in supervision on the part of the colleges is naturally matched by reduced interest among potential master teachers in the schools. Even so, selection is normally poor anyway: weak teachers, often in their own first or second year of teaching, volunteer and are assigned student teachers.

Methods courses

The methods course represents a somewhat more complicated situation. It, too, is affected by the changed direction of the schools of education. More importantly, the status of methods courses is also affected by a long-standing battle over the role of applied science in the college. Engineering and the medical professions have fought this battle and reached an accommodation between theory and practice. (Who would, after all, consider going to a dentist trained only in theory?) Anyone who has taken a laboratory course in the sciences has met at least a vestige of this battle: twice as much work for half as much credit. It is this argument against applications that is usually raised to justify reducing the time devoted to methods instruction.

Every student has his own picture of a good methods teacher in mathematics: a person knowledgeable in his subject, a skilled teacher himself, a leader who can help students gain personal insights into their own teaching problems and assume responsibility for meeting their problems. He needs to be current, to know the new texts and materials in mathematics. He needs to be close to the local elementary schools so that he can draw upon them for support of many kinds. To fulfill these and other demands, the methods teacher must bring strong qualities to his assignment. But most of these qualities, if not directly opposed to what will work to his advantage in his college, are at best tangential to his advantage.

Often the kind of person who would fill the role well finds the educational requirements of the appointment unrelated to his interests, and sees as diverting the educational research he would be required to undertake in order to maintain or improve his position. If he does choose to battle the status quo, and is not blocked by the professional requirements, he soon finds other problems. The small budgetary allotments necessary to support good methods teaching (textbook sets, lab equipment, a library of current books and materials) are not forthcoming. Maintaining and improving his credentials in both mathematics and education becomes increasingly difficult. He is neither understood nor appreciated, because he preaches organized and prepared teaching, something his co-workers seldom practice themselves.

Thus the current situation makes it increasingly difficult to obtain and to retain quality people in this capacity. Higher demands and lower returns hardly make for a saturated market. Evidence of the kind of faculty concern that this elicits may be seen in this true story: Word circulated among students at an East Coast university that a teacher of reading methods was tragically inadequate. Enrollment in her classes dropped to near zero. So she was switched to teaching mathematics methods!

What is being done to counter this deteriorating development?

Unfortunately, virtually nothing is being done. Some schools are trying new programs and utilizing new techniques. But even some of these go astray. Consider a program at my own university—a program designed to educate elementary teacher specialists. Each teacher trainee is required to include a subject major in his undergraduate program. On the surface, this sounds excellent: Elementary schools will be able to draw upon the particular subject matter depth of each individual teacher to maintain and improve standards. But, in fact, does the program produce this? To date it has produced exactly *no* mathematics specialists! Why? Because the forty hours of mathematics (no CUPM Level I courses available) effectively eliminate such a choice. The teachers gravitate to majors that are less demanding in terms of both time and effort. Consequently, public schools end up with specialists in sociology, economics, occasionally psychology, but never mathematics, seldom science, English, history. A basically good program is vitiated by the inflexibility of its own design.

Many mathematicians and some educators feel that the CUPM (Committee on the Undergraduate Program in Mathematics of the Mathematical Association of America) recommendations for course work for elementary school teachers are the answer to teacher-training problems. They are not. They do provide a course-work base upon which to build the education program, but often they do that program an unintended disservice. CUPM-type courses are substituted for methods courses by administrators or, worse still, by teachers who find lecturing on mathematics easier than helping teachers to develop effective mathematical pedagogy.

Most disheartening of all is the fact that the National Council of Teachers of Mathematics is doing little or nothing to battle the problem. This failure has encouraged CUPM to move in to fill some of the gap, but the results of their work are not yet available.

What then should be done?

Most important, we must recognize the critical nature of this problem. Something must be done and done soon. Here are some actions the reader may wish to take:

1. Compare the comments made in this article with the *current* posture of your own institution or alma mater. If you find that the same situation pertains, make known your opposition to specific aspects of the programs through organized teacher and alumni groups.

2. Support at the same time any activities of schools of education that appear to upgrade teacher education. In this commentary, negative aspects of such programs have been stressed. Some of the positive measures undertaken to strengthen teacher education in a meaningful way include: well-designed internship programs (cooperative school-college programs where the teacher trainee is a paid school staff member with close contact maintained with his college faculty); school teaching centers (the college goes to what the students describe as "where it's at," a newer version of "where the action is"); microteaching techniques and videotaping of both stu-

dents and master teachers for methods classes; strong cooperative relationships established between college staff members and classroom teacher organizations; in-service education centers patterned after the Japanese model (teachers return to college for week-long seminars on current teaching and content problems with the local school supplying substitutes for their classes).

3. Obtain the support of college mathematics staff members to help reorient the administration of professional education. Mathematicians can not only exert pressure upon educators within the university, but also they can aid in staffing schools of education. By recommending strong subject-matter-oriented educators and by supporting joint appointments when possible, they can often tip the balance when there is a choice between hiring a new teacher educator or a new academician. At the same time, ask mathematicians to examine with care their attitude toward mathematics pedagogy, and even toward secondary school teaching. Many college and university mathematicians are good teachers whose depth of understanding of their field (and the awe in which their students hold them) carry them over rough spots in their own instruction. They fail to recognize, however, either the difficulties of teaching at lower levels, where students are not as self-sufficient even as college students, or of the value of elementary courses. The point they fail to appreciate is that students have the opportunity to attack problems mathematically at every school level, and that the better we equip teachers to help them to do so, the better will be the student results at all levels. Volumes of content, poorly taught, do not contribute to this goal.

4. Urge the National Council of Teachers of Mathematics, through its president, to form a committee to explore the national extent of the situation described here, and to determine ways the Council can act (possibly in cooperation with other groups such as science and English teachers) to head off and perhaps turn its direction.

5. Attack, by any avenues open to you, the distressing problems that contribute to this trend. In effect this means asking schools of education either to be concerned with the basic problem of education, classroom teaching, or to remove the word "education" from their letterheads; that is, to fish or cut bait!

The Current Status of Preservice Elementary Teacher-Education Programs

By **John A. Dossey**

In the October 1979 *Arithmetic Teacher*, James Fey (1979) reported on the current status of elementary school mathematics teaching as viewed through the results of a set of studies funded by the National Science Foundation (NSF). This review touched on the content, instructional styles, media, and problems that characterize contemporary K–6 mathematics programs. This article will examine the results of a similar national study of mathematics programs for preservice elemen-

John Dossey is professor of mathematics at Illinois State University where he is involved in teaching mathematics and mathematics education courses at both the undergraduate and graduate levels. He has served on NCTM's Commission on the Education of Teachers of Mathematics and on the MAA's CUPM Teacher Training Panel.

tary teachers which was conducted in 1977 by the NCTM's Commission on the Education of Teachers of Mathematics (CETM). The combination of the results from both studies gives an accurate picture of the status of K–6 mathematics education today.

The CETM study was undertaken in order to establish baseline data for future teacher-education studies and in an attempt to describe the typical preservice elementary school teacher-education program in mathematics education. To meet these goals, a comprehensive questionnaire was developed and mailed to all colleges and universities in the United States having teacher education programs approved by the National Council on the Accreditation of Teacher Education (NCATE) and to 48 Canadian colleges and universities. The overall return

rate was 53.7 percent and it reflected the balance between the size and nature of the schools to which questionnaires were mailed. There was no reason to doubt the validity of the findings on the basis of a bias in the returns.

Results

The main topics sampled by the questionnaire were the mathematics content component, the methods component, and the clinical experiences component of the programs. In addition, the form requested information on admission criteria, teacher backgrounds, majors in mathematics for elementary school teachers, and other related topics. A complete set of the questions and the raw data from the responses are available in an ERIC report (Sherrill 1978).

Content

The nature and scope of the mathematics courses recommended for preservice elementary school teachers has been the subject of discussion by two major groups in the past two decades: Committee on the Undergraduate Program in Mathematics (CUPM), 1971 and Cambridge Conference on Teacher Training, 1967. The results of the CETM survey indicate that the recommendations on the form that these courses should follow and the content they should cover has had some impact, but they have not changed the major instructional patterns a great deal from two decades ago.

Ninety-two percent of the responding colleges indicated that they require their preservice elementary teachers to take at least one course in mathematics content. The distribution of credit hours in mathematics content required by the responding institutions is shown in table 1. After adjusting this data to compensate for differences in course hours reported for universities employing quarter systems (24 percent) and those with semester systems (72 percent), the results indicate that the majority of institutions require that the preservice elementary school teacher take one or two 3-semester-hour courses in mathematics. Only 34 percent of the institutions required a second course in content of their preservice elementary school teachers.

The 1971 CUPM recommendations had stressed the need for 12 semester hours of mathematics content encompassing the basic concepts of number systems (whole through complex), informal geometry, algebraic structures, topics from measurement, and probability and statistics. These suggestions were heavily reflected in the topics included in the preservice programs, but not to the depth suggested by CUPM. The topics emphasized in the required course(s) of the responding institutions were as shown in table 2. Although the data show that programs of the CUPM-level type have not been achieved either in hours or in breadth, the content courses do reflect a broadened view of mathematics and in general follow the topics suggested by CUPM.

The analysis of the course names and the textbooks used indicated that the most frequent course titles were "Mathematics for Elementary Teachers," "Structure of Number Systems," and "Fundamental Concepts of Mathematics." The most often cited textbooks used were Wheeler's *Modern Mathematics: An Elementary Approach;* Meserve and Sobel's *Contemporary Mathematics;* and Dubisch's *Basic Concepts of Mathematics for Elementary Teachers.* An examination of these texts, or their equivalents, will provide additional insight into the nature and level of mathematics content studied by most preservice elementary school teachers.

Major in elementary mathematics

Seventy-four percent of the responding institutions had programs which allowed preservice elementary school teachers to obtain either a major or a concentration (resource area, specialization, and so on) in mathematics. In 90 percent of these institutions the number of yearly graduates of such programs did not exceed ten in number.

The additional mathematics content courses contained in the programs of study followed by these majors in elementary mathematics were selected, in order, from "Calculus," "Algebra," "Mathematics for Elementary Teachers," "Probability and Statistics," and "Geometry for Elementary Teachers." The majority of the preservice teachers electing a major in mathematics tended to take three additional mathematics courses beyond the general required content course described above.

An analysis of these findings indicates that beyond the first course, most elementary school preservice mathematics majors tend to follow courses similar to those designed for preservice secondary school teachers rather than those that focus on the content of K–6 programs. The in-depth look at topics from informal geometry and probability are not studied from the vantage point of the situations in which they appear in the K–6 curriculum nor are they studied to the extent suggested by CUPM.

Table 1
Hours of Mathematics Required of All Preservice Elementary School Teachers

Required hours	Percent of schools requiring
0	8.2
1	0.7
2	0.4
3	16.7
4	11.0
5	2.5
6	29.8
7	1.1
8	6.7
9	11.0
10	6.0
Greater than 10	5.9

Table 2
Topics Emphasized in Preservice Mathematics Courses

Topics	Percent of programs emphasizing
Numeration, whole numbers, fractions, decimals	89
Sets, rational numbers, integers, field properties	81
Shapes, properties of plane figures, constructions	68
Topics from probability and statistics	37
Coordinate geometry	18
Calculator mathematics	11
Computer mathematics	5

Table 3
Hours of Mathematics Methods Required of All Preservice Elementary School Teachers

Required hours	Percent of schools requiring
0	10.6
1	1.8
2	19.1
3	51.1
4	7.4
5	3.5
6	2.5
Greater than 6	4.0

Methods courses

The survey results indicated that 90 percent of the institutions had at least one methods course required of all preservice elementary school teachers. The credit hours of required methods in mathematics in the various programs is shown in table 3.

A sampling of the curricula for these required methods courses focused on three major components: special projects or requirements, methodological topics, and use of manipulative materials. In the first category, the following were the most often employed special assignments or projects: a resource idea or activity file (67 percent); summaries of journal articles (60 percent); an original mathematics game (58 percent); an original mathematics discovery activity (53 percent); a self-contained learning package (39 percent); and a critical textbook analysis (25 percent).

These project requirements were accompanied by classroom emphasis on manipulative materials or mathematics laboratories (91 percent); lesson planning (77 percent); learning theories (71 percent); exposure to NCTM (70 percent) or state (40 percent) materials and activities; and test-item construction (45 percent). Two items receiving far less emphasis were the role and use of hand calculators (26 percent) and computers (7 percent).

These last two items cause some concern, especially when viewed in light of Bitter's (1978) findings on the future shape and requirements of mathematics education programs. Although the content and methods courses have changed with time, the question of their preparation of teachers for the future still looms large.

Other topics receiving moderate to extensive emphasis in the methods courses were individualization of instruction (77 percent); special problems of the slow or gifted (60 percent), and dealing with special education students (37 percent). The emerging problems of urban education (21 percent) and career education in mathematics (19 percent) should probably grow in emphasis as the milieu of contemporary teaching situations and the mathematics information explosion has its impact on both the K–6 curriculum and preservice teacher education programs.

In the area of exposure to manipulative materials, the most frequently used materials were the geoboard (79 percent), place-value materials—multibase blocks, chip-trading materials, and so on (76 percent), geometric shapes—geoblocks, pattern blocks, tangrams (76 percent), colored rods (76 percent), counting materials (76 percent), attribute materials (70 percent),

Table 4
Clinical Experiences in Preservice Elementary Mathematics Education

Experience	Percent of programs requiring	Average number of hours in programs requiring
Observation	76	14
1-to-1 tutoring	62	26
Small group instruction	64	13

and computational games (61 percent). Again, calculators (36 percent) and computers (11 percent) received a low order of emphasis.

Clinical experiences

The clinical (practicum) experiences associated with these preservice programs were divided among observation, one-to-one tutoring, and small-group instruction. The percent of programs requiring each and the average number of hours associated with each activity is shown in table 4. The data for the number of hours are somewhat hard to categorize as some of the programs had field-based curricula, while other programs were very clearly campus oriented. The average hours were derived from the data given by the institutions having each of the three forms of clinical activities listed in table 4.

A question inquiring for information about the previous professional experiences of the mathematics content and methods teachers in teacher-education programs found that 51 percent of the methods teachers had never taught in grades K–6 and 43 percent of the content teachers had not taught in grades K–6. The average methods teacher with K–6 experience had had her or his most recent elementary teaching experience 5 years previously. For the content teachers, the average length of time since their last elementary classroom teaching experience was 8 years.

Summary

The consistency of the responses and the comparison of the findings with other sources indicate that the foregoing data accurately reflect the current status of preservice elementary school teacher-education programs in mathematics. These findings, when compared with suggested program criteria and the current K–6 teaching situation and curricula, are somewhat discouraging. In the area of content, the materials covered are too brief and not commensurate with the needs of the K–6 classroom teacher. The number of hours of required mathematics needs to be increased. In the area of methods, the number of hours is again inadequate for the preparation of effective teachers of elementary school mathematics. There do not appear to be either enough offerings or enough variety of mathematics education courses to effectively educate a teacher to deal with the special problems of teaching mathematics to the broad range of children found in the typical classrooms in grades K–6.

This data does establish a baseline for mathematics educators to measure the future progress in preservice mathematics teacher-education. They also provide a guide to the areas of concern and need. It remains for teacher educators to act now to improve the state of mathematics content and methods for preservice elementary school teachers.

Note. The author was a member of NCTM's Commission on the Education of Teachers of Mathematics when this report was prepared. The results reported reflect the work of many CETM members over the past four years.

References

Bitter, Gary C. *Mathematics Education in the Future.* Unpublished manuscript, Arizona State University, 1978.

Cambridge Conference on Teacher Training. *Goals for Mathematical Education of Elementary School Teachers.* Boston: Houghton Mifflin, 1967.

Committee on the Undergraduate Program in Mathematics. *Course Guides for the Training of Teachers of Elementary School Mathematics.* Berkeley: CUPM, 1971

Fey, James T. "Mathematics Teaching Today: Perspectives from Three National Surveys." *Arithmetic Teacher* 27 (October 1979):10–14.

Sherrill, James M. *NCTM Commission on the Education of Teachers of Mathematics Survey of Preservice Elementary Mathematics Teacher Education Programs.* Columbus, Ohio: ERIC (Document Reproduction Service Number ED 162 876), 1978.

Teacher Preparation—
a Coordinated Approach

By **William F. Burger, Lee Jenkins, Margaret L. Moore, Gary L. Musser,** *and* **Karen Clark Smith**

In the September 1981 issue, Dossey reported the "somewhat discouraging" comparison of recommended program criteria with the current teaching situation and curricula for grades K–6. We at Oregon State University had had similar concerns, and in the 1981 fall term we implemented a revised, comprehensive mathematics preparation component for the elementary school teacher education program. Based on the Dossey article, this revised program would rank in the top 5 percent of programs in the country, and we believe it contains the elements necessary to produce an exemplary program.

Some of the recommendations from the past decade influenced the construction of the OSU program:

1. In 1971, the Committee on the Undergraduate Program in Mathematics (CUPM) stressed the need for 18 quarter-hours of mathematics content encompassing the basic concepts of number systems, informal geometry, algebraic structures, topics from measurement, and probability and statistics.

2. During the 1970s the Mathematics Methods Project (MMP) was developed at Indiana University under a grant from the National Science Foundation (NSF). This program inte-

All of the authors are members of the faculty at Oregon State University in Corvallis. William Burger is an assistant professor of mathematics. Lee Jenkins is chairman of the Department of Elementary Education. Margaret Moore is an assistant professor in the Department of Science and Mathematics Education. Gary Musser is a professor of mathematics. Karen Smith is a graduate teaching assistant in education.

grated content, methods, and clinical/field experience. Although this funded program was successful in integrating the three components of teacher preparation, it was not widely implemented—perhaps because of the difficulties in cutting across administrative and departmental lines in universities.

3. In 1976 the Oregon Teacher Standards and Practices Commission, after public hearings, mandated that all elementary school teacher preparation programs include at least 12 quarter-hours of mathematics content and 3 quarter-hours of methods of teaching mathematics.

4. In 1979 Leitzel, Schultz, and White at Ohio State University compared three different models for the mathematics preparation of elementary school teachers and made the following recommendations:

(a) Mathematics faculties, education faculties, and public school personnel should join together in upgrading the mathematics program for prospective elementary school teachers.

(b) Instruction in mathematics content and methods should be strongly coordinated, but need not be combined.

(c) School experience should be included in the coordinated content-methods-school experience package.

(d) A mathematics content course should precede a combined content-methods-school experience package.

(e) A significant portion of the content and methods instruction should be activity oriented; moreover, the activity-based learning should lead the non-activity-based instruction whenever both are present.

5. In 1980 the NCTM published a booklet, *An Agenda for Action: Recommendations for School Mathematics of the 1980s.* Three of its recommendations were that—

- "problem solving be the focus of school mathematics in the 1980s;

- basic skills in mathematics be defined to encompass more than computational facility;

- mathematics programs take full advantage of the power of calculators and computers at all grade levels."

6. Dossey mentioned in his article that—

- the number of hours of required mathematics needs to be increased (70% of the surveyed programs required only 6 hours or less),

- the number of hours of required methods courses needs to be increased (82% of the surveyed programs required only 3 hours or less),

- there needs to be a greater variety of mathematics education courses to educate a teacher to deal effectively with the special problems of teaching mathematics.

The Oregon State University program, based on the foregoing recom-

PRESERVICE MATHEMATICS PROGRAMS

mendations, is a coordinated program with the following elements.

1. The program has the following course requirements:

> 12 hours of mathematics (courses specifically designed for elementary school teachers)
>
> 3 hours of mathematics projects coordinated with the first 9 hours of mathematics
>
> 3 hours of methods of teaching mathematics coordinated with three other methods courses and twenty hours per week of school experience
>
> 1 hour of computer literacy
>
> 3 hours of microcomputer/BASIC programming (a second 3-hour course on programming may replace one of the mathematics courses)

All credits are in quarter hours; quarters are 10 weeks in length.

2. School experience in the sophomore year

3. Student teaching, senior year

4. Various elective courses in content and methods.

During their freshman year, students take three 3-hour mathematics courses, which have an associated laboratory component, together with a coordinated 1-hour projects course. The content material is covered on Monday, Wednesday, and Friday; the laboratory is on Tuesday; and the projects course is on Thursday. Topics covered in these courses include problem solving, numeration, whole numbers, computation (mental, written, and electronic), number theory, fractions, decimals and per cent, ratio and proportion, integers, rational and real numbers, probability and statistics, measurement and the metric system, polygons, applied problem solving, geometric constructions, tessellations, motions in the plane, symmetry, congruence, magnification, and similarity. Students are often introduced to topics at the concrete level on Tuesdays, and discuss or develop projects that are appropriate to use in teaching these topics on Thursdays.

During their junior year, students take an elective course entitled "Problem Solving in Mathematics." This course is designed to update and extend their content backgrounds as they learn to solve problems. Specific heuristics a la Polya are studied and particular emphasis is given to using the calculator as a tool for solving problems as well as for learning mathematics.

The coordinated 1-hour projects courses help the students learn how to translate their knowledge of mathematics into classroom practice using an activity-based approach. Each student organizes an idea file for her or his school experience. The manipulative materials that students work with include pattern blocks, Unifix cubes, geoboards, centimeter cubes, Cuisenaire rods, Base-ten blocks, tangrams, fraction tiles, and a variety of other manipulative materials, both commercial and teacher-made.

The elementary mathematics methods component is thirty classroom hours of instruction coordinated with an equal amount of time in language arts, science, and social studies (for example, having students write an ABCs book of geometry, measuring in science experience, and the history of mathematics). The topics covered in the mathematics content courses mentioned in an earlier paragraph are also addressed with an eye towards how they can be taught effectively.

The emphasis in the first microcomputer course is on problem solving through programming in BASIC. The programming commands introduced include PRINT, GOTO, LET, IF-THEN, FOR-NEXT, GOSUB, string and numeric variable manipulation, one-dimensional arrays, graphics, and a number of APPLE II commands. Throughout the course, students are introduced to modification and utilization of computers in teaching mathematics and other subjects. Students are introduced to preprogrammed problem-solving software, games, simulations, tutorials, and drill and practice examples. Students design and program a concept-formation project to be used with other activities. A second elective course involving more advanced programming in BASIC is also offered.

A wide variety of electives is available to students wishing to take more mathematics, computer science, or methods courses. These include 1-credit and 2-credit short courses in methods during the summer, 3-credit inservice courses in "Topics in Mathematics Education" during the academic year and summer term, additional field experience, a course in which students learn to develop software and to teach programming to intermediate-grade students, as well as traditional mathematics or computer-science courses.

A student who successfully completes a mathematics component such as the one described should be not only well equipped to teach for today, but also well prepared for the substantial changes in store for the curriculum of the future. Best of all, this coordinated approach can be implemented on virtually any campus as part of an elementary school teacher preparation program.

References

Committee on the Undergraduate Program in Mathematics. *Course Guides for the Training of Teachers of Elementary School Mathematics.* Berkeley: CUPM, 1971.

Dossey, John A. "The Current Status of Preservice Elementary Teacher-Education Programs." *Arithmetic Teacher* 29 (September 1981):24–26.

Leitzel, Joan R., James E. Schultz, Arthur L. A. White. "A Comparison of Alternatives and an Implementation of a Program in the Mathematics Preparation of Elementary School Teachers. The Ohio State University. (Final Report to the National Science Foundation) 1979.

National Council of Teachers of Mathematics. *An Agenda for Action: Recommendations for School Mathematics of the 1980s.* Reston: The Council, 1980.

Polya, George. *How to Solve It.* New York: Doubleday and Company, Inc., 1957. ♥

Teacher Education

Preservice Elementary Mathematics Education: A Complete Program

By **John A. Dossey**
Illinois State University, Normal, IL 61761

The concern over the shortage of mathematics and science teachers has once again brought into focus the critical role played by elementary school teachers in the development of children's mathematical abilities. Although many groups making suggestions for improvement lament the mathematical preparation these teachers receive, none have described their conception of a total program of content, methods, and the related clinical experiences and practicum.

The past few years have seen the revision of recommendations for the mathematical education of teachers at all levels, K–12. The most prominent of these are the NCTM's own *Guidelines for the Preparation of Teachers of Mathematics* (1981) and the Mathematical Association of America's *Recommendations on the Mathematical Preparation of Teachers* (1983). The first document lists competencies that teachers at various levels should have upon entering the teaching profession. The second document gives a detailed set of course syllabi for the mathematics courses that are recommended in the NCTM's suggestions. However, neither provides a model for the entire scope of the preservice program for K–6 teachers. Such a model must consider the three interacting dimensions of content; methods; and clinical experiences, or practicum.

The Content Dimension

The content dimension of the program should be based on a minimal prerequisite of three years of secondary school mathematics, consisting of two years of algebra and one year of geometry. The coursework in this part of the program should consist of a six-semester-hour sequence covering the fundamental concepts of the elementary school mathematics curriculum and a three-semester-hour course in topics drawn from geometry and measurement. All these courses should be taught from an advanced viewpoint, emphasizing the development of the central concepts, the major principles, and the critical skills. The discussion of the content should focus on the growth of the core ideas and how they are interrelated. This analysis should also consider the ties between the material studied and the K–6 mathematics curriculum. This connection can perhaps be developed best through the addition of a laboratory-based session to the regular classroom period. Such a session gives the preservice teachers the opportunity to have hands-on work with manipulative materials and models, to discuss the relative merits of various teaching aids, and to "fine tune" their own conceptions of the content being studied. This laboratory session, as well as the regular classroom portion of the class, should be used to help develop the preservice teacher's skills in exploring numerical and spatial situations and in employing problem-solving strategies.

As a result of these studies, the preservice elementary teachers should develop the ability to—

1. identify and develop problems related to the child's environment and illustrate the application of problem-solving strategies built on the child's knowledge of mathematical concepts, facts, skills, and principles;

2. develop and model prenumeration concepts (attributes, classification, ordering, patterns, and sets) using techniques appropriate for various developmental levels;

3. illustrate and explain number and numeration concepts (cardinal and ordinal numbers, place value, and representations of number);

4. develop and explain the usual algorithms for the four basic operations within different number systems (whole-real) and illustrate these operations using models and thinking strategies appropriate for the elementary grades;

5. relate the properties of number systems and their operations to the mastery of basic facts and the execution of the standard algorithms;

6. relate the basic concepts and skills of ratio, proportion, and percentage and illustrate these concepts with the child's experiences;

7. identify and employ examples of simple geometric solids and shapes and the properties relating them to the child's experience and surroundings;

8. use standard and nonstandard units to measure length, perimeter, volume, capacity, mass, weight, time, temperature, and angles;

9. estimate the results of numerical calculations or the measurement

PRESERVICE MATHEMATICS PROGRAMS

25

of a given quantity or object;

10. test the reasonableness of a given explanation or procedure against known facts or prior experience;

11. illustrate the fundamental concepts of probability, inference, and the testing of hypotheses using everyday experiences;

12. solve simple problems involving the reporting of data (measures of central tendency, dispersion, expectation, and prediction) through experimentation, record keeping and graphing, or calculation;

13. make appropriate use of calculators and computers in problem solving and in exploring and developing mathematical concepts;

14. describe the historical and cultural significance of some of the major mathematical concepts and principles studied in the K–6 mathematics program.

These objectives should be central to the content dimension of the preservice mathematics education program for K–6 teachers. In addition, preservice teachers planning to develop a specialty area in mathematics should take additional coursework in the application of computers in the elementary mathematics classroom; advanced topics in number theory, algebra for elementary teachers, and problem solving; and a course in either the history of mathematics or applications of elementary mathematics.

The Methods Dimension

The methods dimension of the program should be carefully sequenced with the content courses or, perhaps, even integrated with them. In either approach, the coursework must be designed to build strong links among the content, teaching methods, and the elementary school mathematics curriculum. In addition, this portion of the preservice teacher's mathematics education must foster the development of teaching strategies appropriate for learners of varying ability and motivational levels. Finally, the methods course must prepare the student with the skills for adapting to curricular changes. Experience indicates that

a minimum of five semester hours is needed to achieve these methodological goals and to monitor the related clinical experience.

One appropriate organizational form for the methods course might focus on the following topics: the elementary mathematics curriculum; types of mathematical content (concepts, facts, skills, and principles) and related teaching strategies; development and fostering the learning of facts, algorithmic techniques, and problem-solving skills; diagnostic and remedial procedures for common elementary mathematics classroom learning problems; and the planning, delivery, and evaluation of mathematics instruction in the K–6 classroom.

The development of these topics should support preservice teachers in the mastery of the earlier content goals, as well as prepare them to—

1. outline the major content areas of the elementary school mathematics curriculum and specify which of these areas are developed at each grade level;

2. describe the relationship among concrete, semiconcrete, and symbolic representations of mathematical concepts and relationships in terms of classroom materials and activities;

3. employ the basic strategies for developing concepts, learning facts, noting patterns (principles), mastering skills (algorithms), and solving problems;

4. develop a plan for instruction that encompasses the overall goals of review, content development, supervised practice, homework, and evaluation;

5. provide instruction in a fashion that shows clarity of instruction, economy of time use, constant attention to the focus of the lesson, appropriate classroom-management skills, and concern for the individual learner;

6. model a positive attitude toward mathematical situations in both mathematics class and other areas of the school curriculum;

7. diagnose a child's level of mathematical development and offer in-

struction to meet the needs noted.

The meeting of these goals and their supporting activities will assist in the development of the preservice teacher's effectiveness in the K–6 mathematics classroom. The melding of these skills with the content mastered in the content courses described earlier will prepare the preservice teacher for the last phase of the mathematics education program—the clinical experiences, or practicum, dimension.

The Clinical Experiences, or Practicum, Dimension

The clinical experiences contained in the preservice mathematics education program for preservice K–6 teachers provide a capstone to the study of content and methods. It is in these activities that the goals of both areas unite to form an effective, and competent, teacher of elementary school mathematics.

These clinical experiences must be carefully integrated with the various topics in the methods dimension of the preservice program. As such, they must be sequenced from structured observations to full class instruction. At the same time, they must be allowed to grow in both duration and level of responsibility.

Initial experiences might involve the observation of selected students during concept-learning or problem-solving situations. These activities provide the basis for discussion of related teaching actions and methods of meeting individual needs. They also assist in describing effective methods of classroom organization and management. The observation phase might be followed by assisting the classroom teacher in the final stages of a lesson. This assistance might involve working with a small group of students as they complete a structured list of questions or exercises. It might also involve the supervision of an activity worksheet when manipulative materials are employed.

As the students gain in experience, the planning of the lessons should gradually shift from the instructor to the student. As the student grows in confidence, the transition of this re-

PREPARING ELEMENTARY SCHOOL MATHEMATICS TEACHERS

sponsibility becomes necessary and appropriate.

The experiences in the clinical dimension of the preservice teacher's program should include opportunities for whole-class instruction; small-group work; teaching of concepts, facts, algorithmic techniques, and problem-solving skills; and working with students of varying ability levels.

Summary

The successful integration of these three aspects of a complete program of mathematics education for the preservice teacher of elementary school mathematics is a task that requires a great deal of planning on the part of a faculty. In many institutions, it requires cooperation across collegiate and departmental lines. In others, it requires the development of new courses and approaches to traditional tasks. In all institutions, the development of such programs requires attention to detail and hard work so that future generations will not lament over the mathematical education of their children's teachers.

References

Mathematical Association of America. *Recommendations on the Mathematical Preparation of Teachers*. Washington, D.C.: Mathematical Association of America, 1983.

National Council of Teachers of Mathematics. *Guidelines for the Preparation of Teachers of Mathematics*. Reston, Va.: The Council, 1981. ◗

The Teacher Educator's Dilemma: Is More Better?

By **Barbara Berman** and **Fredda J. Friederwitzer**

The current need for increased mathematics education at all levels has received many headlines and much attention from Congress, educators, and the general public. The National Commission on Excellence in Education recommended a greater number of mathematics courses for high school students and more stringent criteria for selection of mathematics teachers. The critics assert that emphasis must be placed on content, with rigorous instruction beginning at the elementary school level.

Institutions of teacher education are faulted for lowering admission standards and for devoting too much time to educational "methods" courses and not enough to subject matter. However, teacher educators must *not* unthinkingly jump on the new bandwagon of stricter standards and intensive content. We must accept the criticism that is due but temper it with the knowledge and understanding that our own experiences have yielded.

Successful educators know that what is taught (i.e., content) depends on the approach used to teach it (i.e., methodology)! We cannot abandon methodology totally in pursuit of upgraded content. In our opinion, problems in the training of prospective elementary school teachers in the U.S. have been created by this very same emphasis operating in reverse!

Barbara Berman and Fredda Friederwitzer are codirectors of Project SITE, a National Diffusion Network mathematics in-service program for elementary school teachers offered through Educational Support Systems, 446 Travis Avenue, Staten Island, NY 10314.

Teacher education institutions have, on the whole, emphasized methodology at the expense of content. Elementary school teachers have spent many hours in methods courses but have not received extensive preparation in mathematics (nor in science, social studies, or language arts) as part of their undergraduate programs. Instead, they may have taken one or two elective courses in the subject, frequently courses that are unrelated and out of sequence. In some instances, they have no college background *at all* in mathematics (Berman 1981). Their knowledge of the subject matter then depends on their own high school preparation, which, if we accept the rampant criticism of high school educators, was none too thorough itself.

We agree emphatically that the mathematics content of teacher-training courses (in-service as well as preservice programs) must be expanded. However, the methodology needed to teach this content to children must be presented as well. The solution is to combine content *and* methodology for that topic. Research has repeatedly shown that teachers teach not only *as* they were taught (Fuson 1975) but also *what* they were taught (Goodlad 1983). Teacher educators, therefore, must consistently model the kinds of behavior in their own courses that they expect their students to employ as classroom teachers!

The past two decades have witnessed an explosion in all types of educational research and, in recent years, the implementation of new research-based instructional programs for children. It is time for these kinds of approaches to be applied to teacher education programs as well. Educational and psychological researchers have accumulated knowledge about adult development, learning styles, and effective training practices. Content-oriented staff-development programs based on this knowledge offer promise in solving the dilemma that teacher education now faces.

References

Berman, B. "An Investigation of the Efficacy of an In-service Program Based on the Multiplier Effect on the Achievement of Elementary School Children." Ph.D. diss., Rutgers University, 1981.

Fuson, Karen. "The Effects on Preservice Elementary Teachers of Learning Mathematics and Means of Teaching Mathematics through the Active Manipulation of Materials." *Journal for Research in Mathematics Education* 6 (January 1975):51–63.

Goodlad, John I. *A Place Called School.* New York: McGraw-Hill Book Co., 1983. ▌

The Mathematics Education of Primary-Grade Teachers

By **Joan P. Isenberg** and **Carol J. Altizer-Tuning**

Mathematics is often viewed as difficult subject matter by students preparing to be primary-grade teachers (kindergarten through grade 2). Many of them need to develop competence and confidence in mathematics in order to teach it in a meaningful and appropriate way. Providing for the preservice mathematics education of primary-grade teachers must be the joint responsibility of both the mathematics educator and the teacher educator.

As school districts continue to express concern about basic skills, minimum competencies, and declining achievement in mathematics, teacher educators must reexamine their programs to ensure the preparation of the most competent primary-grade teachers of mathematics. Attention to the *Guidelines for the Preparation of Teachers of Mathematics* (NCTM 1981) and *An Agenda for Action: Recommendations for School Mathematics of the 1980s* (NCTM 1980) are critical components in this reexamination process. Against this background, we will discuss the general professional preparation of competent primary teachers, as well as the mathematical content and methods used in that preparation.

Joan Isenberg teaches courses in early childhood education and reading and language arts at early childhood levels at George Mason University in Fairfax, VA 22030. She is an assistant professor of education. Carol Altizer-Tuning teaches mathematics methods courses for elementary education majors and supervises student teachers at Longwood College, Farmville, VA 23901.

General Professional Preparation

The general professional preparation of primary-grade teachers must address the development of young children and include cognitive as well as physical, social, and moral development. Preservice teachers must pay attention to how learning fits into the total development of the child. Recent research in human development indicates that individual differences in learning styles, language and thought, and the nature of the teacher/child relationship are key dimensions that relate to children's learning (Stevenson 1975). An in-depth understanding of the dynamics of such processes forms the basis of sound teaching practices. The works of Piaget, Bruner, Gagné, and Dienes have greatly influenced our understanding of how children learn. Such an understanding must precede the implementation of an appropriate mathematics program.

Knowledge of how children think combined with a skill in using that knowledge helps children develop healthy attitudes toward learning mathematics. Curriculum theorists such as Taba (1962) recommend that classroom practices teach children how to think for themselves. The classroom becomes more than a fact-finding place; it becomes a forum for thinking.

Despite the importance of this literature on young children, classroom teachers report that their chief source of knowledge about teaching is their personal experience. They feel that teacher education programs are not helpful, and they avoid education principles, theory, and research findings (Howsam 1981). This pervasive attitude greatly hampers attempts to upgrade education in general and mathematics education in particular. Better preparation in mathematics education at the college level is one way to provide better classroom teaching of mathematics.

Preparing kindergarten teachers in mathematics education is often neglected in the professional program. Most kindergarten teachers can be certified with an endorsement in either elementary or early childhood education, and the two programs are often not the same. In general, an elementary-level endorsement to teach focuses on the elementary grades (kindergarten through grade 6) with training to teach kindergarten absorbed into the whole range of grades. Endorsement in early childhood education focuses on the years from prekindergarten through grade 3. Differences in the mathematics preparation of kindergarten teachers exist because of the differences between the two kinds of certification.

According to Castle (1978), the approach to educating the elementary school teacher emphasizes mathematics as a distinct discipline and a separate area of the curriculum, whereas the approach to educating the early childhood teacher focuses on the *child* as an area of study and encourages future teachers to help children learn mathematical concepts. The results from these two different approaches are contrasted in the classroom. Often teachers from a program for elementary teachers do not appreciate the significance of the preoperational activities that young children need; teachers from a program for early childhood education often do not know when or how to move children from play and manipulative activities to elementary mathematics. Kindergarten children can develop important mathematical concepts, and their teachers can provide activities that are important for their later growth in mathematics. Too often this opportu-

nity is easily overlooked.

All preservice primary education majors need time to work with kindergarten children in mathematics. Rather than perceiving the kindergarten year only as training and drill in preparation for first grade, preservice teachers who work with kindergarten children come to understand what kinds of mathematics are appropriate for kindergarten to learn. In addition, all prospective primary-grade teachers need time to observe, participate in, and react to the total primary mathematics program so that they can be effectively prepared in the mathematics that they will be teaching.

The classroom is a forum for thinking.

Overall goals

The following considerations are offered for improving the mathematics education of future primary-grade teachers, kindergarten through second grade:

1. *Provide opportunities for field-based mathematical experiences at each primary-grade level prior to student teaching.* Students need opportunities to observe, plan, and interact with children at all developmental levels in the area of mathematics.

2. *Give students experience with the materials they will use in teaching.* If the use of concrete materials is important to children's learning and thinking, we must provide experiences with such materials in classes for teachers. If students have not used the materials during their own education, they will be less likely to use them in their own classrooms. Similarly, individual and small-group instruction in the form of learning laboratories or centers may be effectively used in methods courses.

3. *Rely on the strengths of your*

teaching faculty. The general educator and the mathematics educator must work together to equip students with techniques for teaching mathematics in the primary grades.

4. *Encourage maximum communication among teacher educators in specialized content areas.* Helping students identify the commonalities of theory increases their ability to view learning as a unified whole rather than as separate and distinct bits of information. In turn, mathematics education can be further integrated into the total curriculum and can become a useful tool for practical, everyday living.

5. *Place major emphasis on understanding how children learn and develop mathematical concepts.* Mathematicians, educators, and psychologists have many views on how understanding develops. Although their views are not all similar, they are complementary. There seems to be little disagreement that children should learn mathematics in a progression from the use of concrete materials to semiconcrete materials and finally to abstract statements. Teacher educators have a responsibility to provide experiences for preservice teachers that illustrate various procedures and techniques for teaching for understanding.

6. *Encourage prospective teachers to do their own thinking in the classroom.* Using their own ideas will help prospective teachers help children do the same, and perhaps they will be less dependent on the commercially prepared materials for the mathematics they are to teach.

Nothing positive, however, can happen in the classroom unless the teacher has a firm grasp on the content and methodologies necessary to make mathematics education come alive for children growing up in today's complex and highly technological world. We need teachers who are prepared to educate children for tomorrow.

Mathematics Content

Four assumptions about mathematics content underlie the comments that follow:

1. A knowledge of mathematics is important in achieving economic security and a personally rewarding life.

2. Most students—boy or girl, regardless of race, ethnic background, or socioeconomic status—are capable of learning the mathematics of the elementary and intermediate grades. Many of these students are also capable of learning the algebra and geometry taught at the high school level.

3. Good teachers must know more about a subject than they are required to know to teach their students.

4. The teaching of mathematics in the elementary grades is a critical factor in the future mathematical success of most students.

The study of mathematics begins in the primary grades. Even preschool children develop primitive mathematical concepts from which they can proceed to more refined and complex concepts. It is vitally important that teachers of children of this age be both competent and confident in the mathematics that span these early grades.

What mathematics will primary-grade teachers be teaching? The mathematics of the primary grades can be classified into nine areas.

Prenumeration concepts

Children need to be provided with experiences in which they recognize likenesses and differences of objects, sort objects by attributes, put things in some identifiable order, and find patterns.

Numbers and numeration

Children need to develop a concept of number. They must learn to recognize the numerical quality of sets of objects—the "twoness" of two eyes, two hands, two apples, and so on; the "tenness" of ten fingers, ten toes, ten pennies, and so on. Children must also learn to recognize the numerals associated with each number and the structure of our decimal system of numeration. The understanding of basic mathematical concepts aids the development of the algorithms associated with the operations of addition, subtraction, multiplication, and division. For example, an understanding of place value aids in understanding regrouping in addition and subtraction.

Geometry

Primary-aged children need to be introduced to the simple geometric figures—circle, triangle, square, and rectangle, for example. They should

---

The effective use of calculating devices requires a good "number sense."

---

be aware of the distinction between points or things inside a geometric figure and outside the figure as well as the points on the figure. They should also be able to find examples of "real-life" things shaped like circles, squares, or polygons.

Relations

Young children can learn to distinguish between simple mathematical relations like the following:

$$7 > 3$$
$$2 < 5$$
$$4 + 2 = 6.$$

Operations

After children have developed a concept of number and know the numerals that are associated with the numbers, they can learn operations with numbers. In the primary grades, children become acquainted with addition and subtraction, simple multiplication, and even some notion of division. They must learn algorithms for these operations and, even more important, must learn to recognize the situations in which each operation is used. For example, if Mary and Jason put their money together to buy some candy, what operation do you use to find out how much money they have to spend on candy?

Measurement

Before they use rulers, children need measuring experiences. Teachers can ask questions like the following: Who is the tallest child in the class? How do we find out? Which glass holds the most chocolate milk? How can we tell? They should also measure things with nonstandard units—for instance, Juan's foot, Thuan's foot, Ginny's foot.

Organization and interpretation of data

Even young children can become interested in collecting data about themselves: What color eyes do the children in our class have? What color eyes do most of the students in this class have?

Problem solving

Young children can be encouraged to be good problem solvers even before they have learned to read either words or numerals. Of course, the problems must be appropriate for the children's levels of learning and interests (Castle 1978; Stevenson 1975). Children attack problems with unfettered minds until they have been conditioned to do otherwise. Some of their ideas for solutions may initially be unusual, but children can learn by their own trial and error to refine their guesses and test their solutions.

Children in the primary grades will need to present and test their solutions orally. Their spoken vocabulary may be imprecise, but if teachers listen carefully to children's words and watch their body language, they can usually follow their thinking. Teachers must themselves know enough mathematics to recognize good mathematical concepts and thinking when they see them in a primitive form. And having seen these concepts for what they are, teachers need to know where the children's mathematical concepts fit into the scheme of mathematics so they can build on the concepts the children already have and help the children sharpen their thinking and extend their vocabularies.

Computing devices

Most people see elementary school mathematics as arithmetic—a collection of number facts and operations with numbers. Teaching algorithms for the four fundamental operations becomes an end in itself. The presence in the classroom of devices that can perform the fundamental operations in fractions of seconds raises questions about the nature of mathematics and how it should be taught. Minicalculators and microcomputers will certainly reduce the need for hours of practice on such exercises as long division with four-digit divisors. But the effective use of calculators and computers will require a better "number sense" than ever before. Operators of computing devices need to know what operations are needed and to have some approximate idea about the size of the results. Children need to know when the answers the computing devices give them are reasonable, and they need to understand how to check their calculations. These skills rely on an understanding of elementary mathematics.

Courses for teachers in the primary grades

The NCTM Commission on the Education of Teachers of Mathematics (1981) has made specific course recommendations for teachers in the primary grades. They suggest the following three courses, each equivalent to three hours a week for one semester—a total of nine semester hours:

1. Number systems, from the natural and whole numbers through the rational numbers
2. Informal geometry, including measurement, graphing, and geometrical constructions, and the ideas of similarity and congruency
3. Methods of teaching the mathematics of early childhood and the primary grades, including diagnosis and remediation of children's learning difficulties in mathematics

In recommending these particular courses, the commission assumes that preservice teachers have a high school background of the equivalent of two years of algebra and one year of geometry. If students in a teacher-training program do not have this mathematical background, they will need to take these courses as part of their preparation for teaching.

The commission's recommendations exceed the current requirements in most teacher education programs (Dossey 1981). The standards of the past, however, are inadequate for the present and future decades. The technological developments of the past decade have minimized the need for algorithmic skills at the same time that they have increased the need for sound mathematical thinking in daily life.

Methodological concerns

Teachers of mathematics in the elementary school have a variety of decisions to make and a still wider variety of options to implement those decisions. Teachers begin with a group of students and some general ideas about the sequence of mathematics topics to be taught in elementary schools. Planning to teach the various topics will almost certainly include decisions about things the children must understand, problems they must solve, and computational skills they must acquire.

In the course of teaching mathematical topics, teachers need to—

- use teaching techniques appropriate for developing understanding, problem-solving ability, and computational skill;
- use textbooks and stimulate students to learn by reading the material contained in those books;
- use concrete materials and visual aids when they will contribute to understanding;
- use laboratory activities to provide applications of the mathematics being learned;
- provide remedial instruction for students with learning problems.

Teachers must adapt their instruction to the many different needs, interests, abilities, and learning styles of their students. They also must continuously assess their efforts by using various evaluation procedures and instruments. Evaluation may include observing students as they complete written practice activities or solve problems. It may also include diagnostic testing using locally developed or standardized tests.

Teacher competencies

Although the foregoing discussion represents an oversimplified view of teaching mathematics, the complex nature of teaching is evident. Teachers need to learn to teach for a variety of objectives and to select content and learning activities that will achieve those objectives. They must offer readiness activities that aid the development of concepts as well as promote the memorization of basic facts and the use of mathematics in practical applications.

Teachers must develop many competencies to lead students to desirable achievement levels in mathematics as well as promote positive attitudes toward the subject. These competencies are developed not only through the study of selected topics in preservice methods courses but also in in-service training sessions. Areas of competence should include the following:

1. Knowing how children learn and how their thinking develops

2. Using manipulative materials and visual aids effectively in the development of specific concepts and skills

3. Selecting drill-and-practice activities to develop speed and accuracy in computation

4. Planning and implementing an appropriate problem-solving program that will improve students' ability to use and apply mathematics (Problem solving has been identified by NCTM as the major focus of school mathematics in the 1980s.)

5. Using the potential of calculators and computers in the full range of mathematics instruction

6. Assessing students' progress, diagnosing errors, and implementing remediation

7. Grouping students for effective instruction

8. Adapting techniques for teaching mathematics to exceptional learners

9. Applying strategies to help students read the language of mathematics

Summary

Teacher educators, state department of education personnel, mathematics supervisors, and those currently teaching mathematics in the primary grades must join together to plan courses and in-service training sessions that will give teachers the confidence and competence essential to the effective teaching of mathematics. Any program designed to improve the teaching of mathematics must integrate the general professional preparation, the mathematics content, and the methodological concerns addressed in this article at both the preservice and in-service levels of teacher education. The process is never ending.

References

Castle, K. "Suggestions for Kindergarten Mathematics Teacher Education." In *Mathematics Teacher Education: Critical Issues and Trends*, edited by Douglas Aichele. Washington, D.C.: National Education Association, 1978.

Dossey, John A. "The Current Status of Preservice Elementary Teacher-Education Programs." *Arithmetic Teacher* 29 (September 1981):24–6.

Howsam, R. B. "The Trouble with Teacher Preparation." *Educational Leadership* 39 (November 1981):144–47.

National Council of Teachers of Mathematics. *An Agenda for Action: Recommendations for School Mathematics of the 1980s*. Reston, Va.: The Council, 1980.

National Council of Teachers of Mathematics, Commission on the Education of Teachers of Mathematics. *Guidelines for the Preparation of Teachers of Mathematics*. Reston, Va.: The Council, 1981.

Stevenson, Harold W. "Learning and Cognition." In *Mathematics Learning in Early Childhood*, edited by Joseph N. Payne, pp. 1–14. Thirty-seventh Yearbook of the National Council of Teachers of Mathematics. Reston, Va.: The Council, 1975.

Taba, Hilda. *Curriculum Development*. New York: Harcourt, Brace, Jovanovich, 1962. ◆

Mathematics for Special Education Teacher Trainees

By **Ellen J. Hofeldt** *and* **Larry L. Hofeldt**

College students and university mathematics professors frequently ask, "Do special education majors need detailed training in mathematics?" This article will assert and delineate the necessity of an all-encompassing background for special education teacher trainees. The special education teacher trainee certainly needs no less thorough training in any area than a regular elementary teacher trainee.

In the very first consideration, the regular education curriculum and the special education curriculum must maintain contact and correlation. A completely isolated or totally self-contained circumstance is discouraged among special educators today. The special student is referred from or started in the regular education program; in some cases an eventual or partial return to regular programming is anticipated. Therefore, anything of value to the regular elementary education major should also transfer in potential usefulness to the special education major. Besides, the greatest job potential in a competitive market is for the trainee who is qualified for both regular education plus special education. After all, only an absolute maximum of 12 percent of the school population needs special education.

Of course, there is always the rebuttal, "But I am going to teach . . ., not somebody who needs all this." Rarely will a special education program include coverage for only one disability. Certification standards vary enough to cause assignment to unanticipated positions. As an example, in Wisconsin, teacher certification in mental retardation is blanket for trainable and educable,

encompassing all age groups. Provisional assignments to some other area can also be made on the basis of the retardation training. But even more important, handicaps are not always easily isolated in either diagnostic techniques or in student placements. Furthermore, a single handicapped child may display a multiplicity of academic, social, or emotional inadequacies. A special education major, indeed, cannot really anticipate what he or she is going to teach.

Regardless of what the teaching assignment is, the mandatory philosophy must be, *There is no one right way to teach a mathematics concept to handicapped children*; if one technique does not work, another must be tried. This is perhaps a very, very astounding thought to some teachers, trainees, and professors. Whether we are teaching counting, adding, or dividing, one method simply will not cover everybody. Some avenues are just plain blocked by the nature of the handicap. If, for example, the associative or reasoning channels of the brain are damaged, there may be no other avenue than pure rote memory; a last resort, but yet that very same primitive manner by which the child first responded to his or her name. In the case of damaged neural areas controlling reasoning, "how or why it works" may not be understood, but yet the neural number areas may function adequately. In any objective consideration of various methods, we should also bear in mind that a major learning reversal must be made in mathematics. In our country we read and write words, even numbers, in left to right progression, but we perform the operations of adding, subtracting, and multiplying by right to left progression, doing the units column first, then the tens, then the hundreds. Thus, what is logical to the author of the text or to the teacher may be only confusion to the child.

Let us look specifically at various special education categories and relate

them to the mathematical inclination. A child of normal intelligence may be placed in a learning disabilities special education class because he does not read, yet his mathematics achievement is average or above in assignments void of reading. A child of mildly retarded intelligence may be placed in an educable special education class because he has brain damage to various portions of his brain in various degrees; it could be that the portion controlling numerical operations is totally intact. A child who was placed in an educable class because he appeared retarded by test and achievement may actually be an emotionally disturbed child, not a retarded child. If the teacher of educables gets this child's emotions settled, instruction would suddenly proceed beyond the educable mentally retarded arithmetic level. The teacher will be striving, during moments of positive student behavior, to get enough taught so that the child will demonstrate higher intelligence in retesting. A child might be referred to a class for the emotionally disturbed; the teacher could be teaching a 7-year-old or a 13-year-old genius. Emotionally disturbed children come in all IQs. In either of the two cases just described, a teacher dare not be groping for mathematical techniques; that teacher must be equal to the task at that very instant.

Mainstreaming in Mathematics

Regardless of whether the special class is for learning disabilities, educable mentally retarded, emotionally disturbed, or some other special title, current trends and research dictate that the child so placed be "integrated" or "mainstreamed" into the regular classroom as much as possible. Mainstreaming of a special student could thus occur in mathematics. How could a slow student be expected to look successful in the regular room if his special

Currently a teacher of educable mentally retarded students at the Roosevelt Elementary School in Oshkosh, Wisconsin, Ellen Hofeldt is also certified to teach kindergarten through eighth grade. Larry Hofeldt, an assistant professor of mathematics at the University of Wisconsin—Oshkosh, has taught courses for teacher trainees. He has found, among the special education majors, a special concern about the training in mathematics.

education teacher cannot follow up or help him with what the regular education teacher introduced? Mainstreaming and the special student both fail miserably unless the special teacher can carry on and augment the concept development immediately. Every present or potential special education teacher should inspect the regular education textbooks and visit the regular education classrooms to see how the mainstreamed child must function. The mainstreamed child is supposed to become socially and educationally adept from going to the regular education room of his approximate age group; that mainstreamed child will have to be pretty great to look good and still be a "slow kid," and still tackle the work. The mainstreamed student has a harder and bigger job than the regular students, and the special education student represents the minority in the regular classroom. The special educator is hired to fill the gap. In this context, if mainstreaming is deemed a desirable attribute of a special education program, then the special educator had better be equal to the obligations and outcomes of the practice.

Other Important Aspects

There are two other aspects of special education that are directly related to how much mathematics a teacher trainee needs. One is that special education often involves transient families. The second is that special educators have advocated the use of unit instruction.

Students in special education programs often come from transient families; learning handicaps apparent in a child are also often apparent in one or both parents. The parents, being handicapped in ability or stability, attain little, if any, job security. Very little cumulative information follows the child as the family moves from one school system to another. In some cases, moves have been so numerous as to prevent all teachers from identifying said child's specific achievements. Unless a teacher can accurately interpret and analyze the child's comments, work, and errors, much previous learning might be confused or lost through initiation of a different

approach to the same process. A thorough knowledge of all approaches to a mathematical operation minimizes this possibility and saves much instructional time. Psychologists can give achievement tests and intelligence tests, but the test results and their reports hardly tell what approach to use to teach a specific concept. Also, there is little available in the way of workable published mathematics programs for special education. Prescriptive teaching principles discourage using routine texts and workbooks. The complaint is that a special education referral student has already failed in texts.

Special education students especially need a consistent instructional pattern, not smatterings. The unit approach to education has been advocated by special educators to train and inform the handicapped child in what he must know to live independently and be self-sustaining. However, a unit provides background, motivation, and use of knowledge on a certain topic. Units must utilize and apply basic operations; in a sense, units must be plugged into the child's basic academic pattern. Lasting knowledge in money, time, measures, or distances depends on order, on a consistent concentrated approach—if a child does not know which is bigger, seventy-five or fifty-seven, he literally can do nothing applicable with units on those topics. If a child must learn base, place value, plus digital pattern, all at the same time that he learns to use the unit equipment, he will assuredly be overwhelmed.

Summary

Special education teacher trainees need everything that a regular elementary education teacher trainee needs, plus specialized training. The only segment of special education that would be likely to require less extensive academic background would be that of the severely retarded, trainable individuals. Children with this affliction constitute less than 1 percent of the school-age population. Thus, any special education trainee would best receive as thorough a background in mathematics and other academic areas as any other elementary grade teacher if employment and success in teaching special education are desired. ☐

Attitudes of Preservice Teachers

3

ALTHOUGH research has not clearly established that teachers' attitudes toward mathematics have an influence on their pupils' attitudes, interests, and achievement, popular belief holds that there is a significant relationship. Concern that the attitudes of elementary school teachers are transmitted to their pupils is reflected in the number of articles over the years dealing with the attitudes of preservice teachers.

Four articles appeared in the 1960s. First is Dutton's "Attitude Change of Prospective Elementary School Teachers toward Arithmetic," which identifies some changes in attitude between 1954 and 1962. The article also contains the attitude scale developed by the author. This "Dutton Attitude Scale" was used in the studies reported in the next two articles. Smith, in "Prospective Teachers' Attitudes toward Arithmetic," used the scale to compare attitudes of students in 1964 with those reported by Dutton. The scale was used again in 1966 by Reys and Delon, who report their findings in "Attitudes of Prospective Elementary School Teachers towards Arithmetic." Kane used a different instrument, not obviously designed to assess attitudes toward mathematics, and discusses his results in "Attitudes of Prospective Elementary School Teachers toward Mathematics and Three Other Subject Areas."

Not only has "arithmetic" given way to "mathematics"; the concern about attitudes is more recently expressed as "mathematics anxiety." Bulmahn and Young, in "On the Transmission of Mathematics Anxiety," present information on the attitudes of elementary education majors toward mathematics, noting that some mathematics anxiety exists. Becker responds with her own study and claims in "Mathematics Attitudes of Elementary Education Majors" that attitudes are not as negative as implied by Bulmahn and Young. Kelly and Tomhave investigated the connection between mathematics anxiety and mathematics avoidance, including differences between men and women, and outline their findings in "A Study of Math Anxiety/Math Avoidance in Preservice Elementary Teachers."

This section concludes with proposals for overcoming negative attitudes toward mathematics in Larson's "Techniques for Developing Positive Attitudes in Preservice Teachers."

Attitude change of prospective elementary school teachers toward arithmetic

WILBUR H. DUTTON
University of California, Los Angeles, California
Professor Dutton is a member of the department of education,
University of California at Los Angeles.

The importance of attitudes as one aspect of the teaching and learning of arithmetic has been well established in educational literature. Studies by Dutton (1, 2, 3),* Billig (4), and Bendig and Hughes (5) are representative of methods used to collect data on attitudes toward arithmetic and mathematics. The search for more adequate questionnaire and sampling techniques and factors underlying attitudes toward these subjects continues to be an important area for research.

The problem

The purpose of this study was twofold: (1) to devise new ways to apply an attitude scale prepared by the writer in 1951 and (2) to search for changes in attitudes of prospective elementary school teachers toward arithmetic since 1954.

The instrument used

The attitude scale prepared by the writer in 1951 was used in this study. This scale was prepared according to techniques developed by Thurstone and Chane (6). While the original scale contained 22 items ranging in value from 1.0 (extreme dislike) to 10.5 (extreme liking), the scale used in this study was reduced to 15 items. This was done to shorten the scale and to eliminate duplication of statements with approximately the same scale value.

* Numerals in parentheses refer to the references at the end of this article.

The final scale provided adequate coverage of scale values at each point between 1 (dislike) and 11 (like).

In addition to the attitude statements, five other sections were included in this study. Space was provided for students to: (1) place a circle around one number from 1 to 11 estimating their general feelings toward arithmetic; (2) indicate average grades made in arithmetic while in elementary school; (3) show the grade where attitudes were influenced most; (4) list two things they like about arithmetic; and (5) list two things they dislike about arithmetic. The reliability of the scale, measured by the test-and-retest procedure, was 0.94.

Subjects

The instruments were administered to 127 prospective elementary school teachers enrolled in curriculum classes at the University of California, Los Angeles. Three classes composed of candidates for the general elementary credential, were selected from six similar sections. All students had completed the methods course dealing with the teaching of arithmetic.

Students were either second-semester juniors or first-semester seniors. Their age varied from 20 to 23 years. Since only five male students were enrolled in these classes, the sampling was restricted to women students. Practically all students had taken Algebra I and II and Geometry, in high school. A lower division arithmetic course (Mathematics 38—3 units) is required of all students before entering the teacher education program.

Entrance requirements for the University are based upon a "B" average from an accredited high school and appropriate courses in science, language, mathematics, history, and English. The entering freshman comes from the top 10 to 12 percent of high school graduating classes. The majority of these students had maintained a "C+" average or above while enrolled in the University.

Results

Attitude scale results

Student responses to the attitude scale are shown in Table 1. The value of each item (1.0 dislike to 10.5 like) is shown in column 1. The total number of student responses for each scale item is shown in column 2. The percent of the total for each item is shown in the last column.

Both favorable and unfavorable attitudes were expressed by most students. Seventy-five percent indicated that "sometimes I enjoy the challenge presented by an arithmetic problem." There were 51 percent who liked arithmetic but liked

other subjects just as well. Another 49 percent felt that working with numbers was fun. Forty-six percent felt that arithmetic was interesting, and 36 percent liked arithmetic because it was practical. Only 18 percent thought about arithmetic problems outside of school.

Unfavorable feelings toward arithmetic were expressed by 37 percent of the students by selecting item nine "I don't feel sure of myself in arithmetic." There were 31 percent who were afraid of doing word problems. Sixteen percent avoided arithmetic because they felt "I am not very good with figures," while 10 percent felt that they had never liked arithmetic.

Careful study of Table 1 and the individual student scales revealed that students have ambivalent feelings toward arithmetic. Most students like some aspects of arithmetic and dislike others. For example, 75 percent of the students sometimes enjoy the challenge presented by an arithmetic problem. Students expressing considerable liking for arithmetic also disliked word problems. Slightly over half of the students, 51 percent, liked arithmetic but liked other subjects just as well. Thus the statement made by the writer in 1951 (2), that for a large propor-

Table 1

Responses of 127 prospective elementary school teachers toward arithmetic

Item	Attitude statements	Value of items	Total responses	Percent of total
1. I avoid arithmetic because I am not very good with figures.		3.2	20	16
2. Arithmetic is very interesting.		8.2	59	46
3. I am afraid of doing word problems.		2.0	39	31
4. I have always been afraid of arithmetic.		2.5	10	8
5. Working with numbers is fun.		8.7	63	49
6. I would rather do anything else than do arithmetic.		1.0	6	5
7. I like arithmetic because it is practical.		7.7	46	36
8. I have never liked arithmetic.		1.5	12	10
9. I don't feel sure of myself in arithmetic.		3.7	47	37
10. Sometimes I enjoy the challenge presented by an arithmetic problem.		7.0	95	75
11. I am completely indifferent to arithmetic.		5.2	2	2
12. I think about arithmetic problems outside of school and like to work them out.		9.5	23	18
13. Arithmetic thrills me and I like it better than any other subject.		10.5	4	4
14. I like arithmetic but I like other subjects just as well.		5.6	65	51
15. I never get tired of working with numbers.		9.8	12	10

tion of the group sampled in 1951 unfavorable attitudes toward arithmetic prevailed, must be qualified. Liking or disliking arithmetic is an individual affair. Diagnosing students' feelings about arithmetic and planning corrective measures must be directed toward individual pupils rather than toward extensive group therapy. This same problem will be discussed later in connection with sampling techniques.

Average scores on attitude scale

To obtain an average scale score for each student in the study, all items checked by each student were assigned scale values and an average computed. These average scores (1.0 extreme dislike to 10.0 extreme like) are shown in Table 2. For example, only one student received an average score of 9.0. The median for the sampling was 6.5 with a Q_3 of 7.5 and Q_1 of 4.3.

By computing the average scale value score for each student, some measure of the student's overall attitude toward arithmetic can be shown. This does not preclude the use of the scale as a diagnostic instrument which will show both favorable

Table 2
Average scale value for 127 students in the study

Scale value	Number of students
10.0	0
9.5	0
9.0	1
8.5	7
8.0	5
7.5	19
7.0	26
6.5	12
6.0	9
5.5	5
5.0	9
4.5	5
4.0	4
3.5	8
3.0	5
2.5	8
2.0	3
1.5	1
1.0	0

MEDIAN = 6.5; Q_1 = 4.3;   TOTAL = 127
Q_3 = 7.5;

Table 3
Student judgment of attitudes toward arithmetic

Attitude level	Student judgment
11	7
10	22
9	24
8	23
7	10
6	16
5	10
4	7
3	6
2	1
1	1

MEDIAN = 8.54; Q_1 = 6.42;   TOTAL = 127
Q_3 = 9.88;

and unfavorable attitudes. Combining the number of students who had average scale value scores between 1.0 and 5.5 (48), there are 38 percent of the students in this study who had very unfavorable attitudes toward arithmetic.

Student judgment of individual attitudes toward arithmetic is shown in Table 3. Each student was asked to circle a number between 1 and 11 to show her overall feeling toward arithmetic—(1 representing extreme dislike and 11 representing extreme like). The median was 8.54 with a Q_3 of 9.88 and Q_1 of 6.42. Note that these judgments are considerably higher than those shown in Table 2. These differences can be accounted for by the averaging of both favorable and unfavorable items checked on the scale by each individual to secure the overall value shown in Table 2. In Table 3, students expressed a generalized feeling toward arithmetic.

Attitude change between 1954–1958

The writer was interested in finding out if there had been changes in attitudes of prospective elementary school teachers toward arithmetic since 1954. A comparison was made between student responses on the scale in 1954 and the responses of students in 1962. These comparisons are

Table 4
**Comparison of 289 students' attitudes toward arithmetic in 1954
with a sampling of 127 students in 1962**

Item	Attitude statement	Scale value	Percent of total responses	
			1954	1962
1.	I avoid arithmetic because I am not very good with figures.	3.2	16	16
2.	Arithmetic is very interesting.	8.1	44	46
3.	I am afraid of doing word problems.	2.0	29	31
4.	I have always been afraid of arithmetic.	2.5	16	8
5.	Working with numbers is fun.	8.7	52	49
6.	I would rather do anything else than do arithmetic.	1.0	3	5
7.	I like arithmetic because it is practical.	7.7	38	36
8.	I have never liked arithmetic.	1.5	9	10
9.	I don't feel sure of myself in arithmetic.	3.7	39	37
10.	Sometimes I enjoy the challenge presented by an arithmetic problem.	7.0	72	75
11.	I am completely indifferent to arithmetic.	5.2	—*	2
12.	I think about arithmetic problems outside of school and like to work them out.	9.5	23	18
13.	Arithmetic thrills me and I like it better than any other subject.	10.5	3	4
14.	I like arithmetic but I like other subjects just as well.	5.6	54	51
15.	I never get tired of working with numbers.	9.8	19	10

* Item 11, while developed with other test items in 1951, was not used in the 1954 sampling.

shown in Table 4 for each attitude statement.

Note in Table 4 that the attitudes of students toward arithmetic in 1954 were almost identical with the attitudes of students in the 1962 sampling. Only one item, number 4, shows any marked change —8 percent fewer students seem afraid of arithmetic in 1962 than in 1954. While there seems to be little change in total feelings toward arithmetic, there is some indication of favorable feelings and newer methodology as shown in student responses in the last two sections of this study.

Grade where attitudes were developed

Feelings toward arithmetic are developed in all grades, according to the responses of students. However, the most crucial years seem to be while pupils are enrolled in Grades 4 through 8 as shown in Table 5. This is in agreement with the 1954 study.

Aspects of arithmetic liked or disliked

Students were asked to list two aspects of arithmetic liked most and two aspects

liked least. This technique was used to give equal treatment to favorable and unfavorable feelings. The writer, in reviewing the data collected in 1954 felt that there was the possibility that students with unfavorable feelings toward arithmetic expressed a disproportionate number of unfavorable responses toward arithmetic—leaving the impression that the majority of prospective elementary school teachers disliked this subject.

Looking at Table 6, one discovers eleven aspects of arithmetic liked by prospective

Table 5
**Grade level where attitudes
were developed**

Grade	Number of students
1	1
2	3
3	11
4	23
5	22
6	17
7	13
8	19
9	8
10	10
	127 TOTAL .

Table 6
Aspects of arithmetic liked most by 127 prospective elementary school teachers

Aspects of arithmetic liked most	Number of students
The challenge arithmetic presents	46
Useful, practical applications	40
Definite, precision of concepts	27
Fun, just working with numbers	26
Being able to reason, logical thinking	17
Solving problems	14
Fractions (7), percent (4)	11
Problems we made up	7
Satisfaction in understanding it	6
Games about arithmetic	4
Long problems	3

elementary school teachers. Heading the list are: the challenge arithmetic presents, practical application, precision of concepts, fun with numbers, and logical thinking. New aspects not found in the 1954 study are: making their own problems, satisfactions centering around understanding arithmetic, and the use of games. These factors, as well as the written reactions of some students, seem to indicate the beginning of teaching procedures designed to create favorable attitudes toward arithmetic and a change away from traditional drill procedures. Students in this sampling would have been in Grades 4–8 during the years 1950–1955.

By limiting student free-response statements to two in each category—(favorable and unfavorable feelings) the number of unfavorable responses toward arithmetic was reduced in this study. Five

Table 7
Aspects of arithmetic disliked most by 127 prospective elementary school teachers

Aspects of arithmetic disliked most	Number of students
Difficult, impractical work problems	39
Tiring, boring work	21
Problems too long, long division (4)	15
Drill over processes already known, work sheets	13
Lack of understanding	11
Percent	9
Fractions	6
Write out proof	5

aspects cover the main feelings of students (Table 7): word problems, boring work, long problems, drill, and lack of understanding.

Study of individual responses not shown in the table revealed students with high favorable attitudes toward arithmetic expressing dislike for difficult word problems and boring, repetitive work. There were six students with high positive attitudes toward arithmetic who did not list any negative feelings for arithmetic. Conversely, there were six students with extremely negative attitudes who did not list any positive feelings toward this subject. Another group of fifteen students who had high positive average scores toward arithmetic listed negative feelings toward word problems and boring work.

Grades earned by students and attitudes toward arithmetic

As shown in Table 8, most students in this study earned grades of "A" or "B" in arithmetic while enrolled in elementary school. Only 19 students made "C" grades in arithmetic. This high performance is consistent with the records of this selective group of students who were required to have a "B" average for University entrance.

Interestingly enough, 18 of the students earning "C" grades in arithmetic had very unfavorable attitudes toward this subject. One student who earned a "C" had favorable attitudes toward arithmetic. Eleven of the 18 students had average scores of 2.5 to 3.0 on the attitude scale—see Table 9.

Table 8
Grades earned by 127 prospective elementary school teachers while pupils in elementary school

Grade	Number of students
A	46
B	62
C	19
	127 TOTAL

Table 9
Attitudes of 19 students earning "C" grades in elementary school arithmetic

	Average attitude score	Number of students
	7.5	1
	4.5	1
	4.0	1
	3.5	3
	3.0	4
	2.5	7
	2.0	1
Extreme dislike	1.5	1

Poor attitudes toward arithmetic were also held by 22 students who earned a grade of "B" in this subject while in elementary school two students with "A" grades in arithmetic also had poor attitudes. (See Table 10.)

Conclusions

The main findings of this study pertaining to new ways of using an attitude scale for measuring prospective elementary school teachers' feelings toward arithmetic are:

1 By using the brief scale of fifteen items, students are able to check the statements expressing their feelings toward arithmetic easily and quickly.
2 The scale may be used as a diagnostic instrument to show both favorable and unfavorable feelings toward arithmetic and to indicate students who are especially high or low—the extremes.

Table 10
Attitudes of students earning "B" grades in elementary school arithmetic

Average attitude score	Number of students
5.0	7
4.5	5
4.0	2
3.5	6*
3.0	1
2.5	1*
2.0	2
	24 TOTAL

* Two students with "A" grades in arithmetic.

3 Comparisons of individual or group attitudes may be made by using the percent selecting a particular scale item, or average scores; the latter are influenced by extreme scores, but seem to give a practical overall assessment of attitude toward arithmetic.
4 By the use of student judgments of their attitudes toward arithmetic, guidance can be given to those students who know they have poor attitudes and to students who may have unrealistically high attitudes. Comparisons between student judgments and scale averages should be helpful in this process.
5 Asking students to record grades made in arithmetic was useful in this study. Students making "C" grades had poor attitudes toward arithmetic. On the other hand there was an equally large group of students who made "B" grades who disliked arithmetic almost as much as the "C" students. This factor should be important for prospective elementary school teachers. High achievement does not insure the development of positive attitudes toward arithmetic by all pupils.
6 Students seem to be able to identify the grade or grades in elementary school where attitudes toward arithmetic were formed. Identification of difficult problem solving, complex processes, teaching incidents, and pleasant associations were recalled by students as they tried to remember experiences in learning arithmetic. These judgments have value for the prospective teacher.
7 Limiting the number of "free response" statements to two for each category (like and dislike) eliminated the problem of securing an unusually large number of responses from students representing extreme attitudes toward arithmetic. At the same time, two responses seemed to cover the main feelings of practically all students.

The search for changes in attitudes of prospective elementary school teachers toward arithmetic since 1954 revealed several interesting findings.

1 The attitudes of students toward arithmetic in 1954 were almost identical with attitudes held by students in the 1962 sampling. Two conclusions on this finding seem warranted: (*a*) these students are the product of a type of teaching which was based upon mechanical, drill procedures; (*b*) instruction in the teaching of arithmetic at the university level (even when students identified their attitudes toward arithmetic) did not change the attitudes held by these students. Will teaching experience and in-service educational programs change the attitudes of teachers who have unfavorable feelings toward arithmetic?

2 New applications of the attitude scale and limiting the "free-response" statements have caused the writer to qualify and extend statements made in the 1954 study.

Many students have ambivalent feelings toward arithmetic. The extremes, students with either very positive or very negative attitudes toward arithmetic, are exceptions to the rule.

While it is important to know that 38 percent of the students in this study dislike arithmetic very much, it is also important to know that 24 percent like arithmetic extremely well and 38 percent like arithmetic fairly well—but not enthusiastically. Fifty-one percent of the students like arithmetic, but like other subjects just as well.

3 There was not enough evidence found in this study to indicate any pronounced improvement in the instructional programs of public and private elementary schools directed toward the development of positive attitudes of pupils toward arithmetic. Prospective elementary school teachers reflect attitudes developed in a traditionally oriented arithmetic program.

4 Attitudes toward arithmetic, once developed, are tenaciously held by prospective elementary school teachers. Continued efforts to redirect the negative attitudes of these students into constructive channels have not been very effective. While the best antidote is probably improved teaching in each elementary school grade, continued study should be made of changing negative attitudes toward arithmetic at the university level and through in-service instruction while doing regular classroom teaching.

5 The aspects of arithmetic liked and disliked by prospective elementary school teachers remained approximately the same between 1954 and 1962. Factors liked most were the challenge arithmetic presents, the usefulness of the subject, fun working problems, and the logical aspects of number. Students continued to dislike impractical word problems, boring work, long problems, and drill over work already known.

References

1 DUTTON, W. H. "Attitudes of Junior High School Pupils Toward Arithmetic," *School Review*, LXIV (January, 1956), 18–22.

2 ———. "Attitudes of Prospective Teachers Toward Arithmetic," *Elementary School Journal*, LII (October, 1951), 84–90.

3 ———. "Measuring Attitudes Toward Arithmetic," *Elementary School Journal*, LV (September, 1954), 24–31.

4 BILLIG, H. L. "Student Attitude as a Factor in the Mastery of Commercial Arithmetic," THE MATHEMATICS TEACHER, XXXVII (April, 1944), 170–172.

5 BENDIG, A. W., and HUGHES, J. B. "Student Attitude and Achievement in a Course in Introductory Statistics," *Journal Educational Psychology*, XLV (October, 1954), 268–276.

6 THURSTONE, L. L., and CHANE, E. J. *The Measurement of Attitude*. Chicago: University of Chicago Press, 1948.

Prospective teachers' attitudes toward arithmetic

FRANK SMITH *Stephen F. Austin College, Nacogdoches, Texas*

Dr. Smith is a member of the department of education at Stephen F. Austin College.

Slightly over ten years ago Wilbur H. Dutton* presented the results of a study of the attitudes of prospective teachers toward arithmetic as determined by an objective evaluation instrument. This present study compares the attitudes toward arithmetic of prospective teachers today with those reported by Dutton.

Over a period of five years, Dutton collected statements from prospective teachers about their feelings toward arithmetic. These statements were judged by a set of five criteria, and 45 statements were retained. On the basis of a sorting procedure performed by 120 students, a scale value and a Q value were assigned to each statement. The technique developed by Thurstone was employed.

Twenty-two statements were selected from the 45. These items are the basis of a scale designed to describe attitudes toward arithmetic from an objective viewpoint. The scale also offers students an opportunity to estimate their total general attitude toward arithmetic. The reliability of the scale was measured by the test-retest procedure. The correlation between the two sets of scores, based on an average scale value for the total test for each student, was .94.

Administration of the scale

The scale was administered to 123 students: 116 females and 7 males. No space was provided for the student's name so that each remained anonymous. These students were enrolled in either a course in methods of teaching arithmetic or a seminar in elementary methods. Table 1 presents a breakdown of the classification of the students involved in the study.

Table 1

Classification of 123 Stephen F. Austin State College Students taking the arithmetic attitude scale

Classification	Number of students
Graduate	7
Senior	74
Junior	42
Total	123

Responses to attitude statement

The record of responses for the 123 prospective teachers is shown in Table 2. Statements numbered 13, 16, and 23 were not checked by any student. These statements express extremely favorable and extremely unfavorable attitudes. Statements 2, 11, 15, and 18 all express an unfavorable attitude toward arithmetic. Less than half of the students checked any of these items. Statement 18, "I am afraid of doing word problems," was selected by 46 percent of the students as representative of their feelings.

Statements 8, 10, 14, and 19 all express a favorable attitude toward arithmetic. More than 60 percent of the subjects checked the first three statements mentioned while 46 chose statement 19. Statement 8, "Arithmetic is as important as any other subject," was chosen as

* Wilbur H. Dutton, "Measuring Attitudes Toward Arithmetic," *The Elementary School Journal* (September, 1954), pp. 24–31.

Table 2

Responses of 123 Stephen F. Austin State College Students on arithmetic attitude scale showing percent of endorsement, scale value for each statement, and the results of Dutton's study

Scale value	Statement number	Attitude statement	Frequency of response		Dutton's study— percent
			Number	Percent	
1.0	13	I detest arithmetic and avoid using it at all times.	0	0	3
1.5	20	I have never like arithmetic.	13	11	9
2.0	18	I am afraid of doing word problems.	57	46	29
2.5	11	I have always been afraid of arithmetic.	30	24	16
3.0	22	I can't see much value in arithmetic.	0	0	.3
3.2	15	I avoid arithmetic because I am not very good with figures.	24	20	16
3.3	9	Arithmetic is something you have to do even though it is not enjoyable.	29	24	27
3.7	2	I don't feel sure of myself in arithmetic.	51	41	39
4.6	6	I don't think arithmetic is fun but I always want to do well in it.	32	26	34
5.3	7	I am not enthusiastic about arithmetic but I have no real dislike for it.	35	28	38
5.6	4	I like arithmetic but I like other subjects just as well.	66	54	54
5.9	8	Arithmetic is as important as any other subject.	93	76	83
6.7	14	I enjoy doing problems when I know how to do them well.	79	64	70
7.0	10	Sometimes I enjoy the challenge presented by an arithmetic problem.	76	62	72
7.7	5	I like arithmetic because it is practical.	49	40	38
8.1	19	Arithmetic is very interesting.	56	46	44
8.6	3	I enjoy seeing how rapidly and accurately I can work arithmetic problems.	51	41	52
9.0	12	I would like to spend more time in school working arithmetic.	26	21	16
9.5	1	I think about arithmetic problems outside of school and like to work them out.	42	34	23
9.8	17	I never get tired of working with numbers.	19	15	19
10.4	21	I think arithmetic is the most enjoyable subject I have ever taken.	3	2	4
10.5	16	Arithmetic thrills me; I like it better than any other subject.	0	0	3

Table 3

Self-rating of 123 Stephen F. Austin State College students concerning general feelings toward arithmetic

Scale value	Number of students checking	Percent	Dutton's study— percent
1 (strongly against)	1	.8	0
2	1	.8	1.4
3	2	1.6	2.4
4	4	3.2	8.0
5	6	4.9	8.7
6 (neutral)	16	13.0	11.8
7	12	9.8	12.8
8	19	15.4	19.7
9	24	19.5	16.3
10	12	9.8	11.4
11 (strongly favor)	26	21.1	7.6

representative of the attitudes of 76 percent of the prospective teachers.

Results of student self-rating

All students were given an opportunity to estimate their general feeling toward arithmetic by circling a number on an eleven-point scale ranging from "strongly against" (1) to "strongly favor" (11). Statement 6 was the neutral point. Table 3 presents a record of the students' self-rating.

Fourteen students, representing 11.3 percent of the group, declared themselves to be on the negative side of the scale. A neutral rating was expressed by 16 students or 13 percent of the prospective teachers. Ninety-three students rated themselves on the positive side of the neutral point. This number represents 75.6 percent of the students.

Students' reasons for liking or disliking arithmetic

After completing the rating scale, the subjects were given an opportunity to express their reasons for liking and/or disliking arithmetic. Table 4 presents the reasons listed for liking arithmetic. Mentioned most often were statements such as; "it is interesting and challenging," "necessary for modern living," "practical and

Table 4

Reasons for liking arithmetic given by 123 Stephen F. Austin State College students

Reason	Number of responses
Interesting and challenging	30
Necessary for modern living	22
Practical and useful	10
Because I understand it	6
Gives a feeling of accomplishment	6
It's fun	3
Fascinating	2
Had good teachers	2
It's logical	2
Enjoy working with numbers	1
It conditions the mind	1
It's rewarding	1
	—
	86

Table 5

Reasons for disliking arithmetic given by 123 Stephen F. Austin State College students

Reason	Number of responses
Lack of understanding	15
Written problems	5
Never have done well	5
Poor teaching	4
Always been weak	3
Lack of teacher enthusiasm	3
Was never related to practical situations	2
Afraid of it	2
The way multiplication was taught	1
Working at the blackboard	1
Drill	1
Requires too much thinking	1
Numbers are confusing	1
Exercises used as punishment	1
Takes too much time	1
	—
	46

useful," "because I understand it," and "it gives a feeling of accomplishment."

The most often mentioned reason for disliking arithmetic was a lack of understanding. Also listed were the following: "written problems," "have never done well," "poor teaching," "always been weak," "lack of teacher enthusiasm," "too much busy work," and "afraid of it." Table 5 gives a complete listing of all reasons given by the students for not liking arithmetic.

When the totals for Tables 4 and 5 are combined, the number is greater than the total number of subjects, because some students listed reasons both for liking and disliking the subject.

Estimation of when attitude was developed

The final portion of the instrument gave the prospective teachers an opportunity to estimate when they developed their feelings (good or bad) toward arithmetic. Table 6 presents these data. Slightly less than one-third of the students listed nothing more specific than "in elementary school." Ten students listed the intermediate grades and six named the primary

Table 6

Students' estimation of when they developed their feelings toward arithmetic

Period	Number
Grade 1	3
Grade 2	2
Grade 3	1
Grade 4	3
Grade 5	2
Grade 6	1
Grade 7	2
Grade 8	3
Primary Grades	6
Intermediate Grades	10
All Elementary Grades	39
Junior High	19
Senior High	17
College	7
Always	8
	123

grades as the periods in which their feelings developed. Approximately 30 percent of the students listed secondary school as the period in which their attitudes were developed. College was mentioned by seven students. Eight students professed that they had always had this attitude toward arithmetic.

Findings and conclusions

The findings of this study indicate that data can be collected which are valuable to the preparation of prospective teachers for the elementary school. The results indicate that too many prospective teachers have negative attitudes toward a subject they will be required to teach. Teacher-education programs should strive to change these attitudes.

The data indicate that feelings toward arithmetic are developed in all stages in our educational system. These findings are in agreement with those of Dutton. More than one-half of the students in this study named the elementary school years as the period in which their feelings toward arithmetic developed. This further emphasizes the importance of good teaching in the elementary school.

"Arithmetic is as important as any other subject," "I enjoy doing problems when I know how to work them well," "Sometimes I enjoy the challenge presented by an arithmetic problem," "I like arithmetic but I like other subjects just as well," and "Arithmetic is very interesting" were the favorable attitude statements most often accepted by the subjects.

"I am afraid of doing word problems" was an unfavorable statement often accepted by the prospective teachers. This points out the need for more emphasis on problem solving at all levels in our educational system.

An individual analysis of the scales revealed that many students liked some areas of arithmetic and disliked others. Dutton also found this to be true. Many students may have had too much emphasis placed on some aspect of arithmetic. At the same time, some may never have discovered the relationship that exists between all aspects of mathematics.

Of the 123 students in this study, 88.6 percent declared themselves either neutral or favorable toward arithmetic. Viewed in another way, 3.2 percent were strongly opposed, 3.2 percent were opposed, 17.9 percent were neutral or slightly below neutral, 9.8 percent were slightly in favor, 15.4 percent were favorable, and 50.4 percent were strongly in favor. In Dutton's study, 3.8 percent were strongly opposed, 8.0 percent were opposed, 20.5 percent were neutral or slightly below neutral, 12.8 percent were slightly in favor, 19.7 percent were favorable, and 35.3 percent were strongly in favor.

Most of the students in this study were enrolled in a public school ten years ago, when Dutton conducted his study, and most of them were in the elementary school at that time. Data indicate that the present group is more favorable toward arithmetic than Dutton's group. Regardless of what factor or factors these changes might be attributed to, a concerted effort should be made to improve the attitude of prospective teachers toward this most important subject.

Attitudes of prospective elementary school teachers towards arithmetic

ROBERT E. REYS and FLOYD G. DELON
Columbia, Missouri

*Dr. Reys is assistant professor of education at the University
of Missouri at Columbia. He teaches courses in mathematics
education and educational statistics.*

*Dr. Delon is associate director of research for the South Central
Region Educational Laboratory, on leave from the
University of Missouri at Columbia.*

What are the attitudes toward arithmetic of preservice elementary education majors? At what educational level were these attitudes developed? Are the attitudes of preservice elementary education majors altered by courses in their mathematics preparatory program? Questions such as these are of paramount importance to those engaged in preparation of elementary school teachers.

Earlier investigations in this area have shown a need for improving the attitude toward arithmetic of prospective elementary school teachers [1, 3, and 7].[1] Research has revealed a relationship between learning efficiency and attitude toward arithmetic [5]. Furthermore, there is evidence to support the claim that attitudes toward arithmetic of elementary school teachers are transmitted to their pupils [2].

---

[1] Numerals in brackets indicate references at the end of the article.

During the 1965/66 academic year a research project focused upon the mathematics preparatory program for elementary school teachers was conducted at the University of Missouri at Columbia [6]. This study differed from previous investigations, such as one conducted by Todd [8], in that the focus was upon the overall mathematics preparatory program for elementary education majors rather than a particular course in the curriculum.

One of the purposes of this investigation was the consideration of the opening questions of this article.

Dutton's Attitude Scale [4], which contains fifteen statements reflecting feelings toward arithmetic, was employed in this investigation. This instrument was administered on two occasions to 385 students enrolled in one of the three courses in the mathematics preparatory program for elementary school teachers at the University of Missouri at Columbia.

The mathematics preparatory program was comprised of a content course in mathematics, a lower-division course in methods of teaching arithmetic, and an upper-division course in problems of teaching arithmetic. The members of the sample were requested to complete Dutton's Attitude Scale prior to and upon completion of one of the preceding courses.

Findings

ATTITUDES

Reported in Table 1 is a summary of the statements selected on Dutton's Attitude Scale by the preservice elementary education majors. Each statement has a scale value which indicates the intensity of the respondent's feelings. The scale values range from 1.0 (strong dislike) to 10.5 (strongly in favor) with 6.0 indicating a neutral feeling.

The prospective elementary education majors showed a higher percent of response to five of the seven positive statements on the posttest than on the pretest. Using Mc-

Table 1

Attitude toward arithmetic before and after completion of a course in the mathematics preparatory program for elementary education majors

Statement	Scale value*	Percent Pre-	Percent Post-
1. I avoid arithmetic because I am not very good with figures.	3.2	28.31	22.34†
2. Arithmetic is very interesting.	8.1	34.54	41.82†
3. I am afraid of doing word problems.	2.0	41.58	34.02†
4. I have always been afraid of arithmetic.	2.5	14.80	16.10
5. Working with numbers is fun.	8.7	36.63	43.12
6. I would rather do anything else than do arithmetic.	1.0	10.39	9.87
7. I like arithmetic because it is practical.	7.7	20.78	28.31†
8. I have never liked arithmetic.	1.5	11.17	13.77
9. I don't feel sure of myself in arithmetic.	3.7	52.73	47.53
10. Sometimes I enjoy the challenge presented by an arithmetic problem.	7.0	70.91	72.99
11. I am completely indifferent to arithmetic.	5.2	5.45	5.19
12. I think about arithmetic problems outside of school and like to work them.	9.5	7.79	10.65
13. Arithmetic thrills me and I like it better than any other subject.	10.5	1.56	1.56
14. I like arithmetic but I like other subjects just as well.	5.6	43.90	45.97
15. I never get tired of working with numbers.	9.8	8.05	6.49

* The scale values supplied by Dutton range from 1 (extreme dislike) to 10.5 (highly favorable).

† Difference significant at .05 level of confidence.

Table 2

Five most commonly selected statements reflecting attitudes toward arithmetic

Course	Rank of statement by number*				
	1	2	3	4	5
Mathematics content course					
Pre	9	10	3	1	14
Post	10	9	3	2	14
Methods of teaching arithmetic					
Pre	10	14	9	5	3
Post	10	14	5	2	9
Problems of teaching arithmetic					
Pre	10	5	14	9	2 & 7
Post	10	5	2	14	7

* Statements may be identified by consulting the numbers in Table 1.

Table 3

Self-rating of general attitude toward arithmetic prior to and after completion of a course in the preparatory program

Rating	Percent of response	
	Precourse	Postcourse
1	1.84	1.90
2	2.60	1.04
3	6.23	5.19
4	7.53	7.00
5	11.69	12.99
6	14.28	11.17
7	9.35	15.06
8	14.28	12.47
9	21.30	18.18
10	7.01	10.91
11	3.64	2.08
No response	.43	2.00
Total	100.00	100.00

Nemar's formula for correlated proportions revealed that two of these increases were statistically significant at the .05 level of confidence. Five of the eight negative statements selected by the students indicated a decrease in percent of response from the pretest to the posttest. Two of these percentage decreases were significant at the .05 level.

Although these results lend some support to the contention that the courses in the mathematics preparatory program cultivate a more favorable attitude toward arithmetic, it should be noted that the change in magnitude of some of these percentages was slight. Furthermore, only a small percentage of the sample chose positive statements numbered 12, 13, and 15; whereas a relatively large percentage of these students selected negative statements numbered 9 and 14.

Table 2 contains the five statements most often selected by preservice elementary education majors as reflecting their feelings toward arithmetic. The statements are listed in descending order, with the statements appearing first being checked the greatest number of times. This table reveals that four statements are common in the preinventory and postinventory of each class.

In the mathematics-content course only one positive statement was indicated on the preinventory, while two positive statements were selected on the postinventory. Students in the lower-division methods course selected two positive statements on the precourse scale and three positive statements on the postcourse scale. In the upper-division methods course three of the statements selected on the precourse scale were positive, whereas four statements on the postcourse scale were positive. In each class, from the prestatus to poststatus, one negative statement was replaced by one positive statement.

One question of the attitude scale asked the students to indicate their general feelings toward arithmetic with the same scale as previously described. The responses to this item are listed in Table 3. On the precourse inventory 55.58 percent indicated a favorable attitude toward arithmetic. The postcourse survey revealed that 58.70 percent of the students showed a favorable attitude toward arithmetic. However, the difference between the percent of elementary education majors expressing a favorable attitude toward arithmetic prior to taking a course in the mathematics preparatory program and the percent expressing a favorable attitude after completing the course was not statistically significant.

Table 4

Grades in which the attitudes toward arithmetic of preservice elementary education majors were reported to have developed

Grade	Percent of response
1–2	6.23
3–4	15.32
5–6	19.22
7–8–9	41.55
10–11–12	11.43
Other	2.34
No Response	3.91

WHEN ATTITUDES DEVELOPED

Table 4 reveals the approximate grade level at which the elementary education majors indicated their present attitudes toward arithmetic were developed. The greatest percent of students indicated that their present feelings toward arithmetic were developed in the junior high grades. Less than half of these students reported that their feelings toward arithmetic developed in the elementary grades; of those who did, a majority indicated this development occurred in the intermediate grades rather than the primary grades.

Summary

In recapitulation, these findings indicate that approximately 60 percent of the preservice elementary education majors expressed a favorable attitude toward arithmetic. The mathematics preparatory courses produced some change in the students' attitude toward arithmetic as evidenced by (1) an increase in the percent of response to five positive statements from pretest to posttest, two of these increases being significant at the .05 level of confidence; (2) a decrease in percent of response to five negative statements from the pretest, two of these changes being significant at the .05 level of confidence; and (3) an increase of one positive statement for each of the three classes, when the five most commonly selected statements on the preinventory and postinventory were considered. It has been previously indicated that although some component changes in student attitude were noticeable, the magnitude of overall change was not statistically significant.

It is not surprising that the observed changes in attitude toward arithmetic in only a few months of instruction were small since most of these attitudes were conceived at least five years prior to entering college and perhaps even cultivated through the years. A large scale improvement of these deep-seated feelings might be expected from high quality instruction in a continuous mathematics program for a longer period of time. However, the problem will not be alleviated until favorable attitudes toward arithmetic are fostered throughout elementary, secondary, and collegiate levels.

References

1. BROWN, EDWARD D. "Arithmetical Understandings and Attitudes Toward Arithmetic of Experienced and Inexperienced Teachers." Unpublished doctoral dissertation, University of Nebraska, 1961.

2. BANKS, J. HOUSTON. *Learning and Teaching Arithmetic* (2nd ed.). Boston: Allyn & Bacon, 1964.

3. DUTTON, WILBUR H. "Attitude Change of Prospective Elementary School Teachers Toward Arithmetic," THE ARITHMETIC TEACHER, IX (December 1962), 418–24.

4. ———. "Attitudes of Prospective Teachers Toward Arithmetic," *Elementary School Journal*, LIII (October 1952), 84–90.

5. LYDA, WESLEY J., and MORSE, EVELYN C. "Attitudes, Teaching Methods, and Arithmetic Achievement," THE ARITHMETIC TEACHER, X (March 1963), 136–38.

6. REYS, ROBERT E. "Are Elementary Education Graduates Satisfied with Their Mathematics Preparation?" THE ARITHMETIC TEACHER, XIV (March 1967), 190–93.

7. SMITH, FRANK. "Prospective Teachers' Attitudes Toward Arithmetic," THE ARITHMETIC TEACHER, XI (November 1964), 474–77.

8. TODD, ROBERT M. "A Mathematics Course for Elementary Teachers: Does It Improve Understandings and Attitude?" THE ARITHMETIC TEACHER, XIII (March 1966), 198–202.

Attitudes of prospective elementary school teachers toward mathematics and three other subject areas

ROBERT B. KANE

Purdue University, Lafayette, Indiana

Mr. Kane is associate professor of mathematics and education.

Studies of attitudes toward learning and teaching mathematics have been reported throughout a period of extensive curriculum revision. Dutton (1951, 1954, 1956, 1962, as cited in the bibliography) developed scales to assess attitudes of prospective elementary teachers and children toward arithmetic. When he compared the responses to his scales in 1954 and 1962, he concluded that no significant changes in the attitudes of prospective elementary teachers and children toward arithmetic had occurred between those two years. Smith (1964) later administered the Dutton scales to prospective elementary teachers. When he compared his results with Dutton's of 1954, he found that the 1964 group was more favorably inclined toward arithmetic than the 1954 group. Smith based this conclusion on the results of the subject's self-rated feelings toward arithmetic as indicated on an eleven-point scale from "strongly against" to "strongly in favor." Of Smith's subjects, 88.6 percent declared themselves either neutral or favorable toward arithmetic, compared with 79.5 percent of Dutton's subjects.

How extensively the higher percentage in Smith's study reflected socially acceptable behavior rather than underlying attitudinal dispositions remains obscure. By 1964 knowledge of the revolution in school mathematics could have evoked a substantial Hawthorne effect among prospective elementary school teachers. Smith reported that all of his subjects, at the time the data were collected, were enrolled either in a course in methods of teaching arithmetic or in a seminar in elementary methods.

Other data in the studies by Dutton and Smith suggest the possibility of social bias as a salient variable in 1964. Summarized in Table 1 are the responses in 1954, 1962, and 1964 to those attitude statements reported in all three studies. Note that the responses in 1964 do not exhibit a trend toward more positive attitudes over the earlier responses. The evidence does not seem to be univocal that attitudes in 1964 were more positive than earlier ones.

A new study

THE PROBLEM

The purpose of the study reported here was twofold: (1) to devise a "neutral" instrument on attitudes of prospective elementary school teachers toward mathematics by asking them to respond to items which exhibit no preoccupation with

Table 1
Comparison of attitudes of prospective elementary teachers toward arithmetic in 1954, 1962, and 1964

	Scale value	Percent of total responses		
		1954	1962	1964
I avoid arithmetic because I am not very good with figures.	3.2	16	16	20
Arithmetic is very interesting.	8.1	44	46	44
I am afraid of doing word problems.	2.0	29	31	46
I have always been afraid of arithmetic.	2.5	16	8	24
I like arithmetic because it is practical.	7.7	38	36	40
I have never liked arithmetic.	1.5	9	10	11
I don't feel sure of myself in arithmetic.	3.7	39	37	41
Sometimes I enjoy the challenge presented by an arithmetic problem.	7.0	72	75	62
I think about arithmetic problems outside of school and like to work them out.	9.5	23	18	34
Arithmetic thrills me and I like it better than any other subject.	10.5	3	4	0
I like arithmetic but I like other subjects just as well.	5.6	54	51	54
I never get tired of working with numbers.	9.8	19	10	15

mathematics; (2) to assess the attitudinal structures of prospective elementary school teachers—those who have completed revised mathematics courses and methods courses as a part of their college work—toward mathematics and other subject areas in which they will be teaching.

THE INSTRUMENT

A questionnaire was constructed. The respondent was asked to rank-order the subject areas of English, mathematics, science, and social studies in response to six statements:

1. I enjoyed my work in this field the most in high school.

2. This field was the most worthwhile for me to study in high school.

3. I enjoyed courses in this field the most in college.

4. I learned the most in courses in this field in college.

5. I probably will enjoy teaching this subject the most.

6. I probably will be most competent to teach this subject.

The questionnaire is reproduced as an exhibit at the end of this report.

SUBJECTS

The questionnaire was administered to fifty-eight elementary education majors at Purdue University at the close of their student teaching period. The director of student teaching, who is not affiliated with any of the four subject areas, administered the questionnaire during a two-day conference on student teaching. The questionnaire was introduced as being of interest to the faculty in elementary education. No specific faculty member or field of specialization was linked to the questionnaire.

For the past four years Purdue University has offered a special eight-semester-hour sequence in mathematics for prospective elementary school teachers. Maximum class size is thirty-five, instructors are picked for their competence and interest in mathematics education, and weekly

Table 2
Number of times each possible rank order of subject areas was selected

				K–3 preference						4–6 preference						Total					
				Enjoyed most in high school	Most worthwhile in high school	Enjoyed most in college	Learned most in college	Enjoy teaching	Most competent to teach												
				1	2	3	4	5	6	1	2	3	4	5	6	1	2	3	4	5	6
E	M	S	ss	0	2	0	1	1	1	0	6	0	2	1	1	0	8	0	3	2	2
E	M	ss	S	1	2	0	0	0	5	3	4	1	0	3	2	4	6	1	0	3	7
E	S	M	ss	1	0	0	1	1	0	0	4	0	0	0	1	1	4	0	1	1	1
E	S	ss	M	0	0	1	0	2	1	3	1	3	0	3	2	3	1	4	0	5	3
E	ss	M	S	4	2	1	0	2	2	2	1	2	0	1	5	6	3	3	0	3	7
E	ss	S	M	3	3	3	0	3	4	2	1	1	0	0	0	5	4	4	0	3	4
M	E	S	ss	1	1	1	2	2	3	3	3	1	2	3	6	4	4	2	4	5	9
M	E	ss	S	2	1	0	0	1	2	1	0	4	2	1	2	3	1	4	2	2	4
M	S	E	ss	1	0	0	0	1	1	3	2	2	3	0	3	4	2	2	3	1	4
M	S	ss	E	2	0	1	0	1	0	2	1	2	1	3	2	4	1	3	1	4	2
M	ss	E	S	0	0	1	1	1	0	1	1	2	1	1	3	1	1	3	2	2	3
M	ss	S	E	1	1	1	2	0	0	1	0	1	2	2	0	2	1	2	4	2	0
S	E	M	ss	0	2	2	1	0	0	1	0	3	0	2	0	1	2	5	1	2	0
S	E	ss	M	0	1	0	0	0	0	1	1	1	3	1	1	1	2	1	3	1	1
S	M	E	ss	1	1	3	0	1	1	2	2	1	3	1	2	3	3	4	3	2	3
S	M	ss	E	0	0	0	2	1	1	0	1	1	2	1	0	0	1	1	4	2	1
S	ss	E	M	0	0	2	2	0	0	1	0	3	2	1	0	1	0	5	4	1	0
S	ss	M	E	0	0	2	5	0	0	2	0	0	3	1	1	2	0	2	8	1	1
ss	E	M	S	1	1	1	0	0	0	2	0	3	3	3	0	3	1	4	3	3	0
ss	E	S	M	3	3	2	2	3	2	1	2	1	1	3	2	4	5	3	3	6	4
ss	M	E	S	1	1	0	0	0	0	4	1	0	1	1	2	5	2	0	1	1	2
ss	M	S	E	1	1	1	1	2	0	1	3	1	1	2	0	2	4	2	2	4	0
ss	S	E	M	0	1	0	2	0	0	1	1	1	2	1	0	1	2	1	4	1	0
ss	S	M	E	0	0	1	2	1	0	0	0	1	0	0	0	0	0	2	2	1	0

E = English
M = Mathematics
S = Science
ss = Social Studies

seminars are conducted for the instructional staff. These courses are followed by a three-semester-hour course in methods of teaching elementary school mathematics. The prospective teachers in this study comprised about one half of the first group of students who had completed this reformed mathematics education curriculum at Purdue University.

RESULTS

Responses to the questionnaire are shown in Table 2. The number of times each possible rank order of subject areas was named is listed for each of the six statements. Data are reported according to the respondent's stated preference to teach in either the kindergarten-primary grades or the intermediate grades, and for the total sample. Subject areas are coded.

One way to compress the data is to consider only the first choices for each item. Table 3 summarizes these data.

The existence of relatively favorable attitudes toward mathematics is displayed throughout the entries in Table 3. The most crucial questions are those that focus on the respondent's anticipated preferences among the subject areas as an elementary school teacher. Among the students who would prefer K–3 assignments language arts was the most popular first-place choice, with mathematics second. Among those who hope to teach in Grades 4–6, mathematics placed first most often.

English was ranked first most often in response to more items than any other subject area. In the final column it can be

Table 3
Rank orders of subject areas picked as first choices for each item

	K–3 preference	4–6 preference	Total
1. Enjoyed most in high school	English Mathematics Social Studies Science	English Mathematics Social Studies Science	English Mathematics Social Studies Science
2. Most worthwhile in high school	English Social Studies Science Mathematics	English Mathematics Social Studies Science	English Social Studies Mathematics Science
3. Enjoyed most in college	Science English Social Studies Mathematics	Mathematics English Science Social Studies	Science Mathematics Social Studies English
4. Learned most in college	Science Social Studies Mathematics English	Science Mathematics Social Studies English	Science Mathematics Social Studies English
5. Enjoy teaching	English Mathematics Social Studies Science	Mathematics Social Studies English Science	English Mathematics Social Studies Science
6. Most competent to teach	English Mathematics Social Studies Science	Mathematics English Social Studies Science	English Mathematics Social Studies Science

seen that English was the most popular first choice for four of the six items; science was the least popular first choice for each of these four items. Also, an interesting inversion may be observed. The respondents reversed themselves on the two questions dealing with college courses (3 and 4). On these science was given as the most popular first choice, and English was never given as the first choice. This suggests that while the students considered English to be most enjoyable and worthwhile as a subject area in high school, as well as the area which many of them expect to enjoy most and to be most competent in as elementary school teachers, their collegiate English courses were not as well regarded. In contrast, science was not well regarded on the high school and teaching items, but it was the most popular first choice in response to the college course items.

Tables 4 and 5 display the frequency with which mathematics was selected first and last for each of the six items. Since there are twenty-four distinct rank-order-

ings possible, of which six have mathematics first, the expectation for each cell in the tables is $(0.25)(n)$. For the K–3 group the expectation that mathematics will be selected first is $(0.25)(23)$ or 5.75. For the 4–6 group the expectation is $(0.25)(35)$ or 8.75. For the total group the expectation is 14.5.

Since these expectations assume no bias

Table 4
Mathematics as a first-rank selection, item by item

Preference	Item						Expec-tation
	1	2	3	4	5	6	
K–3	7	3*	4	5	6	6	5.75
4–6	11	7	12	11	10	16*	8.75
Total	18	10	16	16	16	22	14.50

* Indicates a difference from the expectation: $\alpha = 0.05$.

Table 5
Mathematics as a last-rank selection, item by item

Preference	Item						Expec-tation
	1	2	3	4	5	6	
K–3	7	8	8	6	8	7	5.75
4–6	8	6	10	8	9	5	8.75
Total	15	14	18	14	17	12	14.50

toward any subject area on the part of any of the respondents, the results displayed in Table 4 are dramatic. Among the K–3 prospects, where a disposition towards language arts is strong, first choices for mathematics occurred above the expectation for three of the six items. Of these three items, two pertained to self-perceptions about teaching in the elementary school classroom. Among those who preferred to teach in Grades 4–6, only one item drew first choices below the theoretical expectation.

In Table 5, where the last-place votes for mathematics are recorded, we find only one item below theoretical expectation among the responses of the 4–6 group and no items below the theoretical expectation among the K–3 group. It appears that the subjects tend to pick mathematics either first or last. The meaning of the data in Tables 4 and 5 may become clearer by studying Tables 6 and 7.

Table 6
Total first-place selections on all items

Subject	K–3	4–6	Total
E	47	57	104
M	31	67	98
S	28	37	65
ss	33	45	78

Table 7
Total last-place selections on all items

Subject	K–3	4–6	Total
E	30	38	68
M	44	46	90
S	33	62	95
ss	33	66	99

Among those who would prefer a K–3 assignment, English as a first-place selection stands out from the rest. The other subject areas cluster together. Among the 4–6 prospective teachers, mathematics seems to stand out as a first-place choice; English is a respectable second; social studies and science are much less popular. An examination of the last-place selections suggests that the K–3 group picked mathematics last more frequently than any other subject field. The 4–6 group chose science

or social studies last much more frequently than they did mathematics or English.

These data suggest that attitude towards mathematics may be a useful discriminator between a prospective teacher's preference for teaching in the primary grades versus the intermediate grades.

The difference between the two groups, K–3 and 4–6, may also be observed in Tables 8 and 9.

Table 8
Mathematics as a first- or second-rank selection, item by item

Preference	Item						Expectation
	1	2	3	4	5	6	
K–3	11	10	8	9	11	14	11.5
4–6	21	24*	16	20	19	23	17.5
Total	32	34	24	29	30	37	29.0

*Indicates a difference from the expectation: α = 0, 05.

Table 9
Mathematics as a third- or fourth-rank selection, item by item

Preference	Item						Expectation
	1	2	3	4	5	6	
K–3	12	13	15	14	12	9	11.5
4–6	13	11*	19	15	16	12	17.5
Total	25	24	34	29	28	21	39.0

* Indicates a difference from the expectation: α = 0, 05.

CONCLUSIONS

1. Perhaps the most impressive general conclusion that can be drawn from this study is that the attitude of these prospective teachers toward mathematics is relatively high. Mathematics and English (language arts) consistently command more positive attitudes than social studies and science. While this generalization characterizes the responses to the entire set of items, it is strikingly true of the responses to items concerning teaching competence and enjoyment. The differences between the K–3 group and the 4–6 group, while clearly observable, are not such that either social studies or science displaces mathematics in the overall ranking of subject areas.

2. Important differences between the K–3 and the 4–6 groups occur. Among

the 4–6 prospective teachers, mathematics enjoys the highest attitudinal status, exceeding even that of English. Among the K–3 prospects, attitudes toward mathematics do not seem to be univocal. While mathematics, science, and social studies seem to evoke about the same number of first-place selections, mathematics received more last-place selections. It appears that prospective teachers who have relatively unfavorable attitudes toward mathematics tend to prefer teaching assignments in the primary grades, while those who have the most favorable attitudes toward mathematics tend to prefer assignments in the intermediate grades.

3. The conclusions of Dutton and Smith, relative to the tenacity with which once-developed attitudes are held, find some support in the present study. While the respondents gave high rating to collegiate science courses, they rated high school courses low and preferred to teach other subject areas in the elementary school. Similarly, while college English courses were rated down, high school English courses and the prospect of teaching language arts both drew high ratings.

On the other hand, almost all of the prospective teachers in this study left the elementary school between 1954 and 1956, before curriculum revision in mathematics had touched the elementary school. Most of them completed secondary school mathematics courses of the prerevolution sort. Nevertheless, at the time the data were collected, these education majors exhibited relatively positive attitudes toward mathematics. What part of this outcome is due to publicity about "new mathematics" and reformed collegiate curricula in mathematics education is not known.

Summary

The study reported here set out to develop a technique of assessing attitudes of prospective elementary teachers toward mathematics and the teaching of mathematics by using an instrument that was not obviously designed to sample attitudes toward mathematics. The instrument was administered by a neutral person in a neutral setting to avoid bias toward any of the subject areas. Results indicate that the prospective teachers tended to have relatively favorable attitudes toward mathematics, and particularly toward teaching mathematics in the elementary school. Relatively positive attitudes toward mathematics and the desire to teach in the intermediate grades seem to be paired.

Exhibit: The questionnaire

During your high school career you took courses in a number of fields. Four of these fields are listed below.

	(1)	(2)
English		
Mathematics		
Science		
Social Studies (History, Govt., Sociology, Economics, etc.)		

1. Think of ranking the four fields listed above in order from, "I enjoyed my work in this field the most," down to, "I enjoyed my work in this field the least."

In column (1): Write a 1 after the name of the field you enjoyed the most, a 2 after the name of the field next most enjoyable, and so on.

2. Think of ranking the four fields listed above in order from, "This field was the most worthwhile for me to study," down to, "This field was the least worthwhile for me to study." In column (2): Write the numerals 1, 2, 3, and 4 to indicate how you ranked these fields.

During your college career you have taken courses in a number of areas. A few of these areas are listed below.

	(3)	(4)
English (Composition and Literature)		
Mathematics		
Science (Biology, Chemistry, Geology, Physics)		
Social Studies (Economics, Pol. Sci., History, Sociol.)		

3. Think of ranking the four areas listed above in order from, "It was the most enjoyable for me," down to, "It was the least enjoyable for me."

In column (3): Write a 1 after the name of

the area which was the most enjoyable, a 2 after the name of the area which was next most enjoyable, and so on.

4. Think of ranking these four areas in order from, "It was the area in which I learned the most," down to, "It was the area in which I learned the least."

In column (4): Write the numerals 1, 2, 3, and 4 to indicate how you ranked these areas.

An elementary school teacher usually teaches all the academic subject matter to his or her students. Areas generally taught are listed below.

	(5)	(6)
Language Arts (Reading, Writing, Spelling, Grammar and Punctuation, Literature)		
Mathematics		
Science		
Social Studies		

5. Think of ranking these teaching areas in order from, "Probably will be the most enjoyable for me to teach," down to, "Probably will be the least enjoyable for me to teach."

In column (5): Write the numerals 1, 2, 3, and 4, to indicate how you ranked these teaching areas.

6. Think of ranking these teaching areas in order from "I probably will be most competent to teach in this area," down to, "I probably will be least competent to teach in this area."

In column (6): Write the numerals 1, 2, 3, and 4, to indicate how you ranked these teaching areas.

7. Many elementary school teachers develop a preference for working at certain grade levels. Circle the letter which corresponds to the sentence which best describes your present preference.

a) I would prefer to work in the earlier grades, K–3.

b) I would prefer to work in the intermediate grades, 4–6.

c) At this time I have no preference.

References

Dutton, Wilbur H. "Attitudes of Prospective Teachers Toward Arithmetic," *Elementary School Journal,* LII (October 1951), 84–90.

————. "Measuring Attitudes Toward Arithmetic," *Elementary School Journal,* LV (September 1954), 24–31.

————. "Attitudes of Junior High School Pupils Toward Arithmetic," *School Review,* LXIV (January 1956), 18–22.

————. "Attitude Change of Prospective Elementary School Teachers Toward Arithmetic," The Arithmetic Teacher, IX (December 1962), 418–24.

Smith, Frank. "Prospective Teachers' Attitudes Toward Arithmetic," The Arithmetic Teacher, XI (November 1964), 474–77.

On the Transmission of Mathematics Anxiety

By **Barbara J. Bulmahn** *and* **David M. Young**

The nature and quality of a child's early exposure to mathematics is a matter of ongoing concern in our profession. This concern has been amplified recently by the studies relating to "math anxiety" in children and adults (Tobias 1976). At the same time, research continues on the relative significance of biological and environmental factors on mathematical achievement (Fennema 1974, Hilton and Berglund 1974). The conclusions of these studies are in some cases confusing and even conflicting, but nearly all recognize that a person's environment has *some* effect on her or his mathematical ability and interest.

Barbara Bulmahn is an assistant professor in the Deparment of Mathematical Sciences at Indiana University-Purdue University at Fort Wayne. She teaches undergraduate mathematics and statistics courses, including content courses for elementary education majors. She has been a secondary school teacher, too. David Young is an assistant professor in the Department of Psychological Sciences at the same institution. In addition to conducting research and teaching in the clinical area at the university, he is also a clinical psychologist in private practice.

Elementary school teachers are a significant part of any individual's early mathematical environment, yet it appears that for many elementary school teachers mathematics is at best a necessary evil. The following hypothesis is submitted for your consideration:

In general, the kind of person who is drawn to elementary school teaching is not necessarily the kind who enjoys mathematics in the broad sense—from its logical beauty to its real-world applications.

As a matter of fact, these two areas of preference, elementary school teaching and mathematics—may have some inconsistencies between them.

Let us begin by saying that we do not consider elementary education majors in our program to be less "bright" overall than students as a

whole. They are, for the most part, highly motivated and conscientious students. Their motivations and interests are different, however, from the motivations and interests of other groups of majors. It was the observation of these differences over fourteen years of college teaching that suggested this rather modest study of attitudes toward mathematics.

The study consisted of two parts. One was a questionnaire on which most responses were on a number scale. The other was a collection of essays on attitudes toward mathematics that were required of all elementary education students in their first course of a three-course mathematics sequence Over 200 students completed the questionnaire; roughly half were prospective elementary school teachers. The others were in finite mathematics and psychology classes and had majors of many different types. Both sexes were included, but as might be expected, nearly 90 percent of the elementary education majors were women, while in the other groups the division was about even.

Forty items were on the questionnaire, including questions on sex, age, class, major, enjoyment and perception of difficulty of various academic disciplines at each educational level (elementary, secondary, college), mathematics courses taken in high school, level of mathematics anxiety, and implications for their career choices. The questionnaire went beyond the scope of the familiar Aiken scale of mathematics attitude, because of the longer history of these students (Aiken 1972 and 1974).

Most of the significant correlations between the variables were not surprising and would agree with studies reported in the literature. Students seem to "like" those subjects that are "easier" for them (or conversely). Most students had a fairly consistent preference for one discipline or another throughout their educational levels, although the essays traced some "shifts" as influenced by unusually good or bad class experiences. (These conclusions do have to be tempered with the fact that responses relating to elementary-school experience may be distorted somewhat by subsequent events.)

What was especially interesting was the fair agreement in preferences for mathematics *and* science, or for language arts *and* social studies. Correlations between those two categories were insignificant or negative. At the risk of oversimplifying, the temptation is to categorize the students as "humanities types" or "science-mathematics types." Furthermore, on the basis of essay responses, the conclusion would be that the "humanities types" are predominant in the elementary-education program.

In the essays, the students were asked to discuss their mathematical backgrounds (up to, but not including the current course). This included positive and negative feelings about mathematics, personal experiences that may have shaped those attitudes, and the individual's perceptions of how these might influence her or his teaching of the subject. Although essays are not considered to be "respectable" statistical instruments, we feel that in this study the essay responses were as informative to us as the questionnaires. About two hun-

dred essays were collected, and no scaled questions could have accurately reflected the wide range of experiences narrated there.

Unfortunately, the students who said "math has always been my worst subject" had lots of company. The frequency of that response, or milder versions of it, was disturbing. Even though the students were assured that this essay would not be considered in their course grade, this *was* a mathematics course assignment, so the number of negative responses is probably a conservative estimate of the true situation. Many of those students admitted that their career options had been limited by their mathematical abilities. Even among those who claimed to have achieved high grades in elementary-school mathematics, many admitted to a real fear of word problems. Furthermore, there seemed to be a sharp distinction between those who liked high-school geometry and those who found it to be the first real mathematical trauma.

Most alarming was the feeling expressed by many beginning education students that elementary teachers do not really have to be very good at mathematics beyond the basic computations. They seem to have the notion that with the teachers' manual in hand, they have all the mathematics they need to know. (We try to change their minds on that matter in the mathematics sequence.)

What conclusions can be drawn from this situation? There are several questions on which further studies should be done. Are the abilities and qualities of personality that are required of a good elementary mathematics teacher really inconsistent (or unlikely to be possessed by one individual)? Is the person who is interested in mathematics and its solid problem-solving aspects less likely to be interested in elementary education in the first place? If that is the case, then we are confronted with a situation in which a large portion of elementary school teachers have a much greater interest in language arts and social studies than in mathematics and science. What implications does that have for their teaching? Is a teacher's interest in (or fear of) a subject conta-

gious? Is "math anxiety" a communicable disease that is being carried by elementary teachers to generations of the student population? If it is, then finding some way of breaking the cycle is imperative.

On the presumption that "math anxiety" is, at least to some extent, a communicable disease, what solutions to this problem are available to us? Is there a "treatment" that will "cure" mathematics anxiety in the humanities types who are attracted to elementary school education? Can people with a love of mathematics be encouraged to teach young children? What are the other alternatives?

One answer to the last question would be to provide mathematics specialists in all elementary schools. The number of mathematics specialists and their instructional responsibilities would vary from school to school, depending on the size of the school and the professional backgrounds and interests of the teachers on those faculties. Mathematics specialists might teach mathematics in a departmentalized program, serve as a resource person for grade teachers, or have a combination of these responsibilities. The qualifications for an elementary school mathematics specialist would include elementary school teacher training, a strong mathematics background in both high school and college, and a love of mathematics.

It is likely that in most school systems a combination of remedies will be necessary, but the job must be done. The consequences of inaction are too great.

References

Aiken, Lewis R. "Biodata Correlates of Attitudes Toward Mathematics in Three Age and Two Sex Groups." *School Science and Mathematics* 72 (1972): 386–95.

———. "Two Scales of Attitude Toward Mathematics." *Journal for Research in Mathematics Education* 5 (1974): 67–71.

Fennema, Elizabeth. "Sex Differences in Mathematics Learning: Why? *Elementary School Journal* 75 (December 1974): 183–89.

Hilton, Thomas L. and Gosta W. Berglund. "Sex Differences in Mathematics Achievement: A Longitudinal Study." *Journal of Educational Research* 67 (January 1974): 231–37.

Tobias, Sheila. "Math Anxiety. Why Is a Smart Girl Like You Counting on Your Fingers?" *MS* (September 1976): 56–59, 92. ◗

Mathematics Attitudes of Elementary Education Majors

By **Joanne Rossi Becker**

A recent article by Bulmahn and Young (1982) in this journal discussed an important topic, the mathematics anxiety of future elementary school teachers. For a long time many mathematics educators have decried the poor attitudes toward the subject of those who will teach young children. Unfortunately, relatively little data have been published documenting these poor attitudes and anxieties (see NCTM 1982). Perhaps Bulmahn and Young have such data, but they did not present them in their paper. Instead, they discussed the results in rather broad generalities that make it difficult to compare results with those of other samples or to determine if the elementary education majors in their sample were substantially more anxious about mathematics than the non–education majors in their sample. The authors made the problem of mathematics anxiety seem more severe than I judged it to be from my experience teaching mathematics to students majoring in elementary education. I decided to gather data to substantiate or refute that judgment. This article presents data on the mathematics attitudes of elementary education majors and compares them to data obtained from other populations.

Joanne Rossi Becker teaches both mathematics and mathematics education courses at San Jose State University, San Jose, CA 95192. Her interest in measuring attitudes was developed by over ten years of teaching mathematics to elementary education majors.

Procedures

The descriptive data presented here were collected from a sample of eighty-one elementary education majors enrolled in a required mathematics course and seventy-one students enrolled in a general astronomy course at a large state university. The students were administered a revised version of seven of the Fennema-Sherman Mathematics Attitude Scales (Fennema and Sherman 1976). The scales measured confidence in learning mathematics (C), attitude toward success in mathematics (AS), perceptions of the attitudes of teachers toward the student as a learner of mathematics (T), mathematics as a male domain (MD), usefulness of mathematics (U), mathematics anxiety (A), and "effectance" motivation in mathematics (EM). Each original scale has twelve items, half worded positively and half negatively. Because the scales were designed for use with secondary school students, some items not applicable to college students were omitted, leaving a total of seventy-seven items. Although Fennema and Sherman (1976) report a high correlation between confidence and anxiety, both scales were used here because of the particular interest in anxiety. Reliability coefficients on the seven scales ranged from 0.64 to 0.90, which indicates a reasonable consistency in students' responses to each scale.

Table 1 shows a sample item from each to give the reader an idea of what the scale attempts to measure. The items are presented in a Likert-scale format; that is, one responds in one of five ways from "strongly agree" to "strongly disagree." Each response is given a score from 1 to 5 and, on each scale except MD, a weight of 5 is given for the response that would have a positive effect on the learning of mathematics. For the MD scale, the *less* a person stereotypes mathematics, the *higher* their score.

Table 1
Selected Items from the Fennema-Sherman Attitude Scales

Scale	Item
Confidence (C)	I am sure I could do advanced work in mathematics.
Attitude toward Success (AS)	I'd be proud to be the outstanding student in math.
Teacher (T)	My math teachers would encourage me to take all the math I can.
Math as a Male Domain (MD)	Girls who enjoy math are a bit peculiar.
Usefulness (U)	Taking mathematics is a waste of time.
Anxiety (A)	My mind goes blank and I am unable to think clearly when working mathematics.
Effectance Motivation (EM)	I do as little work in math as possible.

Results

Table 2 shows the means and standard deviations for each scale for the two samples. Because the scales varied slightly in number of items, *average* mean responses, rather than mean totals, are shown for each scale to make comparison of scales easier. The sample of elementary education majors was 95 percent female, so no breakdowns by sex are shown.

As table 2 shows, the education students scored lower on the anxiety scale (i.e., were more anxious) than any other attitude scale; they also scored significantly lower than the astronomy students. Note that an average score of 3 would indicate a neutral attitude, since that number is assigned if one neither agrees nor disagrees with an item. Although the mean anxiety score of the education majors was (statistically) significantly different from a neutral score, would one classify this sample as "highly anxious"? A look at some individual items on the anxiety scale might clarify the level of anxiety shown by these students.

For example, 51 percent of the education students agreed or strongly agreed with the following statement: "Mathematics makes me feel uneasy and confused"; 71 percent disagreed or strongly disagreed with the statement "I almost never have gotten shook up during a math test."

It does not seem surprising that students get nervous about mathematics tests. It is more disturbing that half the students are scared of, and uneasy and confused about, mathematics. However, the astronomy students had similar, although not as strong, feelings. Fifty-five percent disagreed or strongly disagreed with the statement "I almost never have gotten shook up during a math test." Forty-five percent agreed or strongly agreed that "mathematics makes me feel uneasy and confused," with nearly equal proportions of students who were majoring in a mathematics-related field (e.g., a mathematical or physical science or engineering) and those who were not responding in this fashion.

Discussion

Although definitions of "highly anxious" can vary, it seems inappropriate to classify this sample of prospective elementary school teachers as having an alarming degree of mathematics anxiety. Also, the education students as a whole were rather positive in their attitudes toward success in mathematics and the usefulness of mathematics (both these scores were higher than those of the astronomy sample), and they did *not* stereotype mathematics as a male domain. Therefore, I would not classify the education students as having very negative attitudes toward mathematics; they have attitudes comparable to those of a more general sample of

students.

In addition, one might compare the average anxiety score of this sample of elementary education majors with the original data obtained by Fennema and Sherman (1976) from a sample of females in the ninth through twelfth grades that included both mathematics and nonmathematics students. In that general sample, the average anxiety score was 3.11. Thus, the education majors were somewhat more anxious than a broad selection of high school students. The education majors also had attitudes on the other scales that were similar to those of other samples of high school mathematics students (Fennema and Sherman 1976; Fennema et al. 1981).

Although these preservice teachers' attitudes were not *very* positive, neither were they as negative as implied by Bulmahn and Young. Certainly it would be preferable for all teachers to be very positive about mathematics. But we may be expecting too much if we want education students to be more positive about mathematics than college students in general. In addition, we should be mindful of research that indicates that teachers' attitudes have fewer effects on students' learning than we might intuitively assume (Begle 1979).

The university from which the sample was taken is a fairly selective one, so these data may not be representative of all elementary education majors. It would be useful if other institutions used similar instruments,

so that we would have some comparable data on attitudes of elementary education majors.

Certainly better prepared mathematics teachers would provide better instruction in mathematics, so I concur with Bulmahn and Young's call for mathematics specialists in the elementary schools. Care should be taken, however, not to blame all the problems in the learning of mathematics by children on their teachers' anxieties about the subject.

References

Begle, Edward G. *Critical Variables in Mathematics Education: Findings from a Survey of the Empirical Literature.* Reston, Va.: National Council of Teachers of Mathematics, 1979.

Bulmahn, Barbara J., and David M. Young. "On the Transmission of Mathematics Anxiety." *Arithmetic Teacher* 30 (November 1982):55–56.

Fennema, Elizabeth, and Julia Sherman. "Fennema-Sherman Mathematics Attitude Scales: Instruments Designed to Measure Attitudes Toward the Learning of Mathematics by Females and Males." *Psychological Documents* (1976):Ms. No. 1225.

Fennema, Elizabeth, Patricia L. Wolleat, Joan D. Pedro, and Ann D. Becker. "Increasing Women's Participation in Mathematics: An Intervention Study." *Journal for Research in Mathematics Education* 12 (January 1981):3–18.

National Council of Teachers of Mathematics. *Resources on Mathematics Anxieties of Teachers.* Reston, Va.: The Council, 1982.

Acknowledgment

The author thanks John Broderick, Department of Physics, Virginia Polytechnic Institute and State University, Blacksburg, VA 24061, for his cooperation in collecting data. ◼

Table 2
Average Mean Responses and Standard Deviations

Scale	Number of items	Average mean response (standard deviation) of education students	Average mean response (standard deviation) of astronomy students
Confidence	12	3.23 (1.00)	3.61 (.80)
Attitude toward Success	10	4.26 (.61)	3.97 (.64)
Teacher	9	3.35 (.76)	3.33 (.58)
Math as a Male Domain	12	4.40 (.66)	4.04 (.67)
Usefulness	10	3.80 (.80)	3.73 (.72)
Anxiety	12	2.81 (.96)	3.22 (.76)
Effectance Motivation	12	3.02 (.87)	3.20 (.75)
All scales	77	3.54 (.63)	3.58 (.51)

A Study of Math Anxiety/Math Avoidance in Preservice Elementary Teachers

by **William P. Kelly** and **William K. Tomhave**

In her 1972 study, Lucy Sells (1978) indicated that 92 percent of the female first-year students in the University of California had such inadequate mathematics preparation that they had effectively closed the door on 70 percent of the career choices available to them. Sell's conclusions and subsequent research on math avoidance were the bases for the research we conducted during the 1980–1981 school year at the University of Minnesota, Morris, a liberal arts college with an enrollment of 1700. This study was directed at documenting math avoidance among female students on the campus.

The results showed that the enrollment percentages for men and women for the lower-level courses such as math concepts, introduction to statistics, college algebra, and precalculus matched almost exactly the ratio of men to women for the college enrollment as a whole. Of 787 students enrolled in these courses, 381, or 48.4 percent, were women. However, the data for the courses required for the preprofessional programs and those leading to a major in mathematics indicated that of 1029 students enrolled, only 311, or 30.2 percent, were women. Thus, there is a strong indication that women at this college are avoiding the mathematics courses

William P. Kelly is chairman of the Department of Education at Regis College in Denver, CO 80221. William K. Tomhave teaches at the University of Minnesota, Morris, MN 56267. They are both involved in in-service and preservice work in mathematics education and have a special interest in math anxiety.

necessary for many professional and technological careers.

Background

Tobias (1978) sees math avoidance as the natural consequence of a set of attitudes that develop as a result of a student's early educational experiences. During these years there is an emphasis on timed tests, right answers, ambiguous vocabulary (words such as *root, plane, altitude*), and difficult word problems. As a result, many students have math anxiety, a fear of failure when they attempt to learn the content and process of mathematics. Although there is insufficient evidence that a causal link exists between anxiety and math avoidance, it is Tobias's opinion that a student who avoids mathematics in high school and college usually manifests a relatively high degree of math anxiety at the same time.

The literature indicates that both men and women can be afflicted by math anxiety, but women apparently suffer more (Burton 1979; Osen 1974; Tobias 1980). However, there is no support for the general belief that females cannot do well in mathematics (Fennema and Sherman 1977); nor is there any pattern of consistent superiority demonstrated by one sex over the other (Fennema 1974). Thus, most of the reasons for women's nonachievement in mathematics must be placed under the heading of *societal expectations*: women are not supposed to do well in mathematics; it is a male domain.

As a result of our study, and challenged by the apparent connection

between math avoidance and math anxiety, we decided to measure the extent to which the math avoiders on this campus were afflicted with math anxiety.

Our plan

Since math avoiders are difficult to identify in groups, we had to choose a sample by using groups of students that we believed would be populated by math avoiders. These groups were freshmen who had had no college preparatory mathematics courses; seniors who had had no college mathematics courses; freshmen who were enrolled in college algebra (and thus had had minimal college preparation in mathematics); and a group of students who were enrolled in a workshop for the math anxious. In addition, we included forty-three elementary education majors—only six had gone beyond college algebra in their mathematics preparation.

In order to assess mathematics anxiety, we chose to administer the Mathematics Anxiety Rating Scale (MARS), a ninety-eight-item self-rating scale. Each item on the scale represents a situation that may arouse anxiety and is designated by the notations *not at all, a little, a fair amount, much,* or *very much.* The examiner scores 1 point for *not at all* through 5 points for *very much* on each item. The sum of the points for the ninety-eight items provides the total score for the instrument, giving a range of scores from 98 through 490 (Richardson and Suinn 1972). Low scores indicate low math anxiety; high scores indicate high math anxiety. A sample of the normative data for the MARS is given in table 1.

We separated the scores of the women elementary education majors from those of the men to determine whether or not being both a woman and an elementary education major would produce a math anxiety rating score higher than those of the other groups. (Our results appear in table 2.)

We noted the following:

1. On the average, the elementary education majors scored higher (230.0) on the MARS than any of the other groups except those in the math anxious workshop (321.6).

2. The elementary education males scored lower (194.0) than any other group.

3. The elementary education females scored higher on the MARS (245.6) than any other group except those in the math anxious workshop.

4. With the exception of the 1980 males (173.1), all elementary education subgroups—1980 females (250.1), 1981 males (218.3), and 1981 females (240.7)—scored higher on the MARS than any of the other groups except the math anxious workshop students.

In order to provide a convenient method of comparing the data from our various groups, we compared our results to the normative data (table 3).

These data suggest that a high proportion of the female elementary education majors are math anxious. The female elementary education majors are the only group that parallels the math anxious group in terms of the proportion who score above the 50th percentile.

If the results of our study are representative of preservice teacher education, then women elementary school teachers, who constitute the majority of elementary school teachers, may be perpetuating math anxiety with young girls in their own classrooms.

Preventive measures

Math anxiety seems to be exhibited by two categories of people: students learning mathematics and some pro-

spective teachers who will be teaching mathematics.

We recommend that prospective teachers who are math anxious receive immediate help in the form of support groups directed by profes-

sional mathematics teachers who understand math anxiety and some of its causes (Fauth and Jacobs 1980). Math anxious teachers should be encouraged to trace the origins of their fears and to work at conquering them

Table 1
Normative Data for MARS

University of Missouri sample: n = 397 mean = 215.38
standard deviation = 65.29

Percentile ranks for given raw scores

Score:	123	165	189	215	228	255	267	325
Percentile:	5	20	40	50	60	75	80	95

(Richardson and Suinn 1972)

Colorado State University sample (two testings—same subjects): n = 119 mean = 187.3, 179.9*
standard deviation = 55.5, 55.9

Percentile ranks for given raw scores

Score:	124, 114	146, 133	178, 169	227, 223	289, 292
Percentile:	10	25	50	75	90

* Second number refers to data from the second testing.
(Suinn et al. 1973)

Table 2
Sample Data for MARS, University of Minnesota, Morris

			Number	Mean
Group 1		Seniors who had no experience in college mathematics	15	198.7
Group 2		Freshmen in college algebra (minimal preparation)	14	195.1
Group 3		College freshmen with no preparatory mathematics	12	208.1
Group 4		Math anxious workshop students	10	321.6
Group 5		Elementary education male	13	194.0
	a)	1980	7	173.1
	b)	1981	6	218.3
Group 6		Elementary education female	30	245.6
	a)	1980	17	250.1
	b)	1981	13	239.8
Group 7		Elementary education combined	43	230.0

Table 3
Comparison of Sample Data to Normative Percentiles—Numbers of Subjects at or above the Given Percentile

Group/percentile	5	20	25	40	50	60	75	80	95	Mean
Norm raw score	123	156	165	189	215	228	255	267	325	215.38
Math anxious (n = 10)	10	10	10	10	10	10	9	9	5	321.6
Elementary education females (1980, n = 17)	17	17	17	16	11	10	9	7	2	250.1
Elementary education females (1981, n = 13)	13	11	10	9	9	8	7	6	1	240.3
Elementary education males (1981, n = 6)	6	6	6	4	4	4	1	0	0	233.4
Freshmen (n = 13)	11	11	11	7	6	3	1	1	0	208.1
Seniors (n = 15)	14	13	11	7	3	3	2	2	1	198.7
College algebra (n = 14)	13	12	11	7	6	3	1	1	0	195.1
Elementary education males (1980, n = 7)	7	3	3	1	0	0	0	0	0	173.1

PREPARING ELEMENTARY SCHOOL MATHEMATICS TEACHERS

through exposure to mathematics concepts and processes in a nonthreatening environment.

Students, however, should be confronted with their math anxiety and its causes as soon as it appears. The anecdotal records of our subjects indicated that most math anxious people can probably trace the cause of their anxiety to some of the following: timed tests, overemphasis on right answers and the right method, working at the blackboard in front of peers, lack of acceptance of nontraditional problem-solving methods, and such negative counseling as, "You won't need this," or "You aren't smart enough to learn this." These conditions may still be prevalent in classrooms, and they can be harmful influences in the development of a positive attitude toward learning mathematics (Fennema 1982).

Further study

We are currently conducting research on the effect of mathematics teachers' attitudes on their students. Although it is only conjecture, we believe that students who are surrounded by confident teachers who are excited and positive about their role in the students' learning process will exhibit fewer symptoms of math anxiety than students whose teachers are themselves anxious, uncomfortable, and negative about teaching mathematics.

References

Burton, Grace M. "Regardless of Sex." *Mathematics Teacher* 72 (April 1979):261–70.

Fauth, Gloria C., and Judith E. Jacobs. "Equity in Mathematics Education: The Education Leader's Role." *Educational Leadership* (March 1980):485–90.

Fennema, Elizabeth. "Mathematics Learning and the Sexes: A Review." *Journal for Research in Mathematics Education* 5 (May 1974):126–39.

———. From an address to the Minnesota Council of Teachers of Mathematics, Minneapolis, 1982.

Fennema, Elizabeth, and J. Sherman. "Sex-related Differences in Mathematics Achievement, Spatial Visualization, and Affective Factors." *American Educational Research Journal* 14 (Winter 1977):51–71.

Osen, L. *Women in Mathematics*. Cambridge: M.I.T. Press, 1974.

Richardson, Frank C., and Richard M. Suinn. "The Mathematics Anxiety Rating Scale: Psychometric Data." *Journal of Counselling Psychology* (November 1972):551–54.

Rocky Mountain Behavioral Sciences Institute. "Mathematics Anxiety Rating Scale." Fort Collins, CO 80651.

Sells, Lucy. "Mathematics: A Critical Filter." *Science Teacher* 45 (February 1978):28–29.

Suinn, Richard M., C. A. Edi, J. Micoletti, and P. R. Spinelli. "The MARS, a Measure of Mathematics Anxiety: Psychometric Data." *Journal of Clinical Psychology* (November 1973):373–74.

Tobias, Sheila. "Math Anxiety: What You Can Do about It." *Today's Education* (September-October 1980):26GS–29GS.

———. *Overcoming Mathematics Anxiety*. Boston: Houghton Mifflin Co., 1978. ☙

Teacher Education
Techniques for Developing Positive Attitudes in Preservice Teachers

By **Carol Novillis Larson**
University of Arizona, Tucson, AZ 85721

Many college seniors entering a course in elementary school mathematics methods admit to having negative attitudes toward mathematics. For most of these future elementary schoolteachers, the course in mathematics methods will be their last encounter with mathematics prior to student teaching. Since attitudes can probably be transmitted from teachers to students, developing a positive attitude toward teaching mathematics is one of the major goals of the course. The course in mathematics methods cannot be just a clinic on math anxiety however; other goals must also be realized. These other goals include (1) attaining knowledge of the curriculum and (2) being able to select and use appropriate teaching aids, instructional strategies, classroom organizational patterns, and diagnostic and evaluative techniques. Nevertheless, no conflict really exists between covering the usual content of a course in mathematics methods and at the same time encouraging positive attitudes toward mathematics and combating math anxiety. A principal ingredient is "to deal with feelings," as well as with ideas and teaching skills. This article describes techniques that can be used to help change mathematical attitudes and that, at the same time, accomplish the other goals of the course.

Initial Discussion of Attitudes

The students need to be told as soon as possible—preferably the first day of class—that developing positive attitudes toward mathematics is a primary goal of the course. This opener should lead to a general discussion of attitudes toward mathematics, math anxiety, and the importance of attitudes in learning and teaching. As the discussion on attitudes progresses, one student invariably admits to knowing about mathematics anxiety firsthand. This first personal story usually serves to open the floodgates. One story of "mathematical misery" is quickly followed by another. Students are often surprised at the number of their classmates who have feelings similar to their own. The discovery that other future teachers also dislike mathematics and are frightened at the thought of teaching it is important in helping them to acknowledge and consciously work on this problem. Now is a good time to recommend that they read *Mind over Math* (Kogelman and Warren 1978) or *Overcoming Math Anxiety* (Tobias 1978) and to encourage them to become more aware of how they feel about the mathematics that they are engaged in daily.

In any discussion of attitudes, another side of the story emerges—some students reveal that they like mathematics and always have. Others describe how one concerned teacher helped them change their attitudes toward mathematics by helping them learn mathematics and by making it fun. This sharing of positive and negative feelings toward mathematics at the very beginning of the course, along with the instructor's acceptance of these feelings, helps to produce an open atmosphere that allows students to express their feelings concerning mathematical topics, activities, assignments, and tests during the entire course.

Working in Small Groups

Students can be organized into small groups for exploring concrete materials and games, finding error patterns, and solving problems. In small groups, students are more likely to participate in an activity and become involved in a discussion. Risk taking is much easier to do in a small group of one's peers. Students are also able to get to know each other better and to develop camaraderie and peer support. Many college students associate small-group discussions with social science and humanities classes; in mathematics classes students expect to be on their own to discover the "one right answer." Thus, working in groups of three or four reduces anxiety because students don't have total responsibility for finding the answer or completing the task.

A Diversity of Approaches

When working on problem solving in small groups, groups are encouraged to explore variations of the problem once they have solved the original problem. When all groups have solved the problem, a member of each group puts its method of solution on the chalkboard. The class discusses the various approaches and the many ways of recording observations and procedures. They are encouraged to express how they feel about what went on in their groups and about sharing their procedures. Many students express surprise at the diversity of approaches. Someone usually exclaims, "Why didn't we do things like this when I was in elementary school?" This awareness of many ways of approaching mathematical problem solving helps alleviate some of their math anxiety, for they had previously held the misconception that there is one and only one way of solving a problem. The combination of working in small *groups* on problem *solving* and sharing *among groups* is an extremely valuable experience. In this type of activity, all the following misconceptions are dealt with simultaneously: (1) Doing mathematics is a solitary endeavor. (2) Each problem has one right answer and one way to get it—most likely by using a complicated equation. (3) Problem solving is hard, and I can't do it. Misconceptions such as these are at the root of math anxiety.

Meaning for Mathematics

Through exploring concrete materials for modeling concepts and algorithms and through participating in demonstration lessons conducted by the instructor, students develop a better understanding of elementary school mathematics. Discovering that every rule that they had previously memorized has a reason, that they can now understand it themselves, and that they can even explain it to someone else helps them to develop self-confidence. It also repudiates another misconception: "I don't have a mathematical mind, so it doesn't matter how hard I try, I just can't understand mathematics." They become aware that they are not only learning mathematics themselves but also learning how to teach it to children.

Sources of Information

Students need to become familiar with both elementary school textbook series and professional publications, especially journals. An assignment that has helped students develop confidence in their ability to read and evaluate material for mathematics instruction entails examining and evaluating one strand in a current elementary school mathematics textbook series and reading six journal articles on the same topic. By following a strand from grade level to grade level in the textbooks, they are able to see the spiral nature of the curriculum, the diversity and complexity of the ideas that comprise a strand, and the repetitiveness of models, explanations, and activities in any given series. This assignment reinforces the importance of not teaching solely from the book. It also gives them a certain amount of reassurance that they don't have the responsibility of knowing every detail of the scope and sequence of the mathematics curriculum—it's there in the textbook. At the same time, the diverse ideas found in such journals as the *Arithmetic Teacher* show them that sources exist for activities to supplement the textbook.

Knowledge, combined with increased self-confidence, helps students realize that they have the ability to teach elementary school mathematics. This combination results in students developing a more positive attitude toward elementary school mathematics. For some students this positive attitude does not extend to mathematics in general—but it is a very important step.

Testing

Students with math anxiety do not like taking tests—they report that they worry about tests, and this worry adds to their math anxiety. An article on techniques for developing positive attitudes would not be complete without mentioning this conflict between reducing anxiety and testing in some students' views. One solution to this problem would be not to use tests as part of the evaluation of the course. However, I have attempted to solve this problem by using criterion-referenced tests and by having test results comprise only 50 to 60 percent of the final grade. The other 40 to 50 percent of the grade is based on assignments outside of class. Although this approach helps, it does not completely solve the problem for some students.

Conclusion

Overall, the techniques described in this article seem to be effective in reducing math anxiety and developing attitudes toward mathematics in preservice teachers. My evidence for this conclusion is the many self-reports from current and former students, such as the following:

1. Ann proudly reported in class one day, "Four of us went out for pizza, and for the first time in my life I volunteered to figure out each person's share of the bill, including the tax."

2. Peggy, approximately thirty years old, told the class the last week of the semester that she had opened a new checking account, her first in five years.

3. Some student teachers report that the first subject that they elect to teach during student teaching is mathematics.

References

Kogelman, Stanley, and Joseph Warren. *Mind over Math*. San Francisco: McGraw-Hill Book Co., 1978.
Tobias, Sheila. *Overcoming Math Anxiety*. New York: W. W. Norton & Co., 1978. ■

Content and
Content/Methods Courses

THE first six articles in this section deal with mathematics content courses for preservice elementary school teachers. A concern of several of the authors is how to relate the mathematics course to the elementary school mathematics curriculum. The final two articles offer suggestions for teaching a combined mathematics content and methods course.

In "A Mathematics Laboratory for Prospective Elementary School Teachers," Fitzgerald describes using laboratory sessions as a modification of the standard lecture format of a mathematics class. Bausell and Moody conducted an experiment to discover whether the topics in a mathematics course could be restructured to make them relevant to topics taught in elementary school and whether preservice teachers could learn mathematics content through teaching it to fourth graders. The results are presented in "Learning through Doing in Teacher Education: A Proposal." In an effort to help students see some relationship between their mathematics course and the elementary school curriculum, Spitzer designed an experience for the study of geometry and describes this in "A Proposal for the Improvement of the Mathematics Training of Elementary School Teachers." Francis shows how a historical perspective can be used to enhance mathematical appreciation in "History of Mathematics in the Training Program for Elementary Teachers."

Leake wrote "Some Reflections on Teaching Mathematics for Elementary School Teachers" to communicate successful ideas to others who teach a course in mathematics content for preservice teachers and to encourage those who teach similar courses to do likewise. Billstein and Lott responded to Leake's call with "More Reflections on Teaching Mathematics for Elementary School Teachers," offering several ideas and suggestions that had worked for them.

In "Training Elementary Mathematics Teachers in a One-Semester Course," Crittenden describes the evolution of a course dealing with mathematics content and teaching methods and involving a laboratory school. Chinn designed her content and methods course to help preservice teachers develop self-confidence and skills for teaching mathematics in activity-oriented ways and outlines the format of her course in "Preparing Teachers for Individualized Teaching of Mathematics."

A mathematics laboratory for prospective elementary school teachers

WILLIAM M. FITZGERALD

Michigan State University, East Lansing, Michigan

The author is an associate professor of mathematics. The innovative project that is reported in his article should be brought to the attention of mathematics educators.

Have you ever had the opportunity to learn mathematics in an individualized, active learning situation? When elementary school teachers are asked this question, fewer than one percent answer affirmatively. In mathematics the learning experience of nearly everyone, including elementary teachers, takes place in a room where all are expected to learn the same thing at the same time, usually as a passive receiver.

Michigan State University requires every prospective elementary school teacher to complete one four-term-hour course on elementary number systems. More than a thousand students enroll in this course each year. To cope with the large number of students, the mathematics department has offered the course in large lecture sections of 300 or more students, meeting four hours each week.

In view of the fact that many elementary school teachers are being told they should "individualize the mathematics" for their students, serious doubt arose about the appropriateness of the format of the course. Many schools are attempting to develop nongraded programs with teaching teams. Projects such as the Madison Project and the Nuffield Project are promoting individualized laboratory approaches to the mathematics curriculum in elementary school. These and other forces are compelling the teacher to provide a more individualized curriculum in mathematics. Might not teachers provide such a curriculum more effectively if they had personally experienced a learning situation in mathematics that was individually oriented and activity-centered?

It was the judgment of the Michigan

State University mathematics department that the standard format of the mathematics class, meeting only in lecture sessions, should be modified to provide a more personalized learning experience. To this end, a laboratory session was instituted. The students continue to attend the large lecture sessions three days each week, but their fourth period is a two-hour laboratory session with a small group of thirty students.

The laboratory sessions are materials-centered. In small groups of one to five, the students work with physical materials for the specific purposes of—

1. Learning the mathematical concepts of the course.

2. Becoming familiar with the materials and how they may be used.

3. Having a real learning experience in mathematics in a student-centered rather than a teacher-centered classroom.

The materials in the laboratory sessions are selected to complement the mathematical ideas presented in the lectures. A few of the materials were developed in the laboratory, but most of them were purchased through commercial sources.

The laboratory instructors are graduate assistants in the mathematics department who were selected because of their previous background in public school teaching and/or their interest and sensitivity toward the kind of atmosphere we are trying to create in the laboratory.

Because of the administrative problems occasioned by the large number of students, the instructors' development of the lab sessions, and evaluation of student performance, the lab sessions have been rather highly structured. Each student has essentially the same experience with the same materials each week. While mathematics laboratory purists may find this format objectionable, we assume that the prospective teacher, when in her own classroom with thirty children, will be able to

provide a more open situation. Orientation to the laboratory sessions is provided by showing the films "Maths Alive" and "I Do and I Understand," which were produced in England.

Below is a listing of the content and related materials used thus far. Each topic represents one two-hour laboratory session.

Content	Materials
1. Numeration systems	Multibase arithmetic blocks
2. Sets and relations on sets	Attribute games and problems
3. Equivalence relations	Colored rods
4. Functions	Madison Project shoe boxes
5. Whole and rational numbers	Colored rods, equation games
6. Number theory	Colored rods and locally made paper and pencil materials
7. Algebraic systems	Colored rods and locally made materials
8. Measurement	Scales, tapes, hypsometers, etc.
9. Probability	Dice, coins, etc.
10. Geometry	Geoboards
11. Topology	Locally made materials

There is time for only nine lab sessions each term, and the above list has been developed during two terms. Topics other than elementary number systems are included in the last four lab sessions, since 80 percent of our students terminate their mathematical studies with this course. It seems important to us that all prospective teachers have some exposure to geometry, probability, and algebra.

Assessment of the effects of the laboratory thus far seems to indicate that the mathematical competence of the students is unchanged. However, the subjective statements from the students reflecting their attitudes are highly positive regarding their laboratory experiences.

The mathematics department will continue the laboratory as an integral part of the course on elementary number systems. In preparation is a laboratory manual which will be available to the students and will allow a variety of professors to teach the course even though they may not be familiar with all of the recently developed materials and techniques for using those materials.

The development of the laboratory was supported by the Educational Development Program of Michigan State University at East Lansing.

Learning through doing
in teacher education: a proposal

R. BARKER BAUSELL and
WILLIAM B. MOODY

Both authors are at the University of Delaware.
Barker Bausell is a research assistant in mathematics education.
William Moody is an associate professor of mathematics education.
He holds a joint appointment in the Department
of Mathematics and the College of Education.

The rationale for teaching mathematics to prospective elementary school teachers is quite straightforward. A teacher must obviously have some knowledge of the discipline's subject matter in order to insure adequate learning on the part of instructed students. This is normally accomplished by requiring the elementary education major to take a prescribed number of courses dealing with concepts deemed relevant to the elementary mathematics curriculum. The problem with this procedure is that the college textbook writer must himself arbitrarily decide which concepts are relevant and which are not. The purpose of the present article is to propose a procedure for teaching mathematics to propective teachers that avoids much of this arbitrariness.

The ultimate test of the relevance of any portion of the elementary education student's mathematics curriculum is whether or not he ever uses the knowledge gained through its study in his subsequent teaching experience. Fortunately, for the textbook writer, the student's teaching experiences almost universally take place after the student has been exposed to curriculum. There is no logical or empirical reason why this ordering could not be altered.

If indeed the chief purpose of teaching mathematics to prospective elementary school teachers is to give them the necessary background to teach mathematics to elementary school children, it follows that those topics taught to prospective teachers should be capable of being restructured in such a way that they are relevant to topics that can be taught to children. If such restructuring is not possible for some topics, then a question arises as to why those topics should be taught to prospective teachers at all.

Such a restructuring would serve at least two worthwhile objectives. In the first place a certain amount of dead wood (from the standpoint of teaching elementary school children) would probably be eliminated. By the same token, some topics may have to be added, if, on careful perusal of the elementary school curriculum, topics are found that have no counterparts in the teacher education curriculum. The second result of restructuring could be a set of materials that would allow the prospective teacher to learn relevant content *while* he teaches it. This would be an especially beneficial circumstance if the old adage is true that the best way to learn a topic is to teach it.

The present authors recently carried out an experiment that was addressed to ascertaining (1) whether mathematical content could be restructured in such a way that it was relevant to students at two such diverse levels, and (2) whether prospective teachers could learn mathematical content through teaching it (Bausell and Moody, 1973).

Two experimental units were constructed for the purposes of the study. One dealt with exponential notation, the other with elementary number theory concepts. Twenty students from a freshmen-level mathematics course volunteered to teach both units to fourth-grade children. These freshmen teachers, along with the rest of the mathematics class, were given both experimental units and told that they were to be tested on the contents of the units in two weeks; the results of this test were to be counted toward their course grades. Each unit contained (1) ten instructional objectives for teaching fourth-grade students, (2) a brief mathematical development for each objective designed specifically to give the undergraduates the necessary mathematical background to teach the objective, and (3) an example of the type of item that the fourth-grade students would be expected to answer as a result of the instruction.

The teacher volunteers were not told which unit they were going to teach until immediately prior to each instructional trial, hence theoretically forcing them to prepare to teach both units both trials. In actuality half of the teacher volunteers were randomly selected to teach exponents both trials and half to teach number theory the first time and exponents the second. It was hypothesized that if the *act* of teaching produced significant learning, then the half teaching number theory on the first trial and exponents on the second would learn more of the number theory unit than their counterparts who did not teach number theory at all. By the same token, the ordering should be reversed for the students teaching exponents twice: they should perform better on the exponent unit. If, on the other hand, there were no learning differences between the two groups of teachers, and if both groups were superior on both units to students who did not teach, then it would have to be concluded that it was not the act of teaching *per se* that caused superior learning but rather the act of *preparing* to teach.

The latter result occurred. Students who taught learned more of both units than students who did not teach, even after initial differences due to mathematical ability and undergraduate grade point averages were taken into consideration. There was no relationship, however, between which units the students taught and their performances on tests based on those units. The tests used to measure the undergraduates' mastery of the experimental materials was based only on the mathematical developments for the two sets of instructional objectives; all answers to all items were found verbatim in these developments.

Some examples of the experimental materials follow. It should be remembered that the instructional objectives were to be used in the instruction of fourth-grade students; the examples were the types of items on which these students were expected to perform as a result of instruction; and the mathematical developments were included as background for the teachers.

Objective 6—Number theory unit. Identify the quotient in a division problem as impossible to determine uniquely when the divisor is zero. Example: *a*) In a division problem the number we divide by can never be the number____, or *b*) $11 \div 0 =$ ____.

Mathematical Development. A problem that is misunderstood by many people involves division by 0. You can *never* divide a number by 0; there is no solution to such a problem.

Consider $8 \div 4 = \underline{?}$. This is solved by asking $\underline{?} \times 4 = 8$ and obtaining 2. However, when attempting to solve $8 \div 0 = \underline{?}$ in a similar fashion we obtain $\underline{?} \times 0 = 8$, and there is *no number* times 0 that will yield 8. The product $\underline{?} \times 0$ will be 0 no matter what number the $\underline{?}$ is replaced by. Therefore, we say there is no answer to a problem like $n \div 0$, when $n \neq 0$.

In the case where $n = 0$, such as $0 \div 0 = \underline{?}$, we have $\underline{?} \times 0 = 0$. Since any number employed as a replacement for $?$ will make this sentence true, we are in the position of not having a unique quotient. We generalize by saying that we never divide by zero.

Other objectives in the number-theory unit were:

Objective 1—Identify when one number is divisible by another number.

Objective 4—Identify whether a number is a prime number.

Objective 4—Exponent unit. Rename a number to the "zero power" as 1. *Example:* $5^0 = \underline{\quad}$

Mathematical Development. By definition $x^0 = 1$ for any counting number x. For example $3^0 = 1$; $20^0 = 1$. It is defined this way to be consistent with other definitions and properties involving exponents.

Objective 6—Exponent unit. Rename the product of two numbers with like bases as the common base with an exponent equal to the sum of the two exponents.

Mathematical development. As a further example of Objective 4, consider the case of x^0 in Objective 6. $x^0 \cdot x^5 = x^{0+5}$, since $0 + 5 = 5$, then $x^{0+5} = x^5$. Thus it must be true that $x^0 = 1$. Since $x^0 \cdot x^5 = 1 \cdot x^5 = x^5$.

Although many teacher educators and elementary mathematics teachers may consider this genre of behavioral objective and corresponding development too restrictive for their purposes, it is the opinion of the present authors that materials corresponding to this basic format can be developed for every topic in the elementary school curriculum. Changes in the types of objectives, or in the explanations of those objectives do not influence the basic concept.

It should be emphasized at this point that the previously discussed experiment did not ascertain whether or not students can learn more through teaching than through traditional classroom instruction in which the faculty member can serve as a viable information source. Even if teaching does not prove to be a learning mechanism superior to traditional classroom instruction, teaching does, however, have several advantages over the latter from a teacher education prospective.

Most obviously, while the student is learning mathematical content through teaching he is also gaining teaching experience. Secondly, not only does the teacher learn, so do his students. Since the present authors have demonstrated (Moody, Bausell, and Jenkins 1973) that students learn more in small groups (less than five) than in larger ones (over 20), it would appear reasonable that if the student teachers instructed children in small groups as a supplement to the regular classroom teacher's instruction (that is, as remediation), then superior overall learning should occur.

In the final analysis, however, there is no guarantee that such a complex integrated process will work. The research so far accomplished is only supportive, not conclusive. A great deal of effort, both in curricula restructuring and evaluation, will be required to institute the program on even a small scale. The present authors believe, however, that the initial prognosis is quite encouraging and suggest that such an ambitious undertaking may be well worth the effort.

References

Moody, W. B., and R. B. Bausell. "Learning Through Teaching." Paper presented at the Annual Meeting of the National Council of Teachers of Mathematics, Houston, 1973.

Moody, W. B., R. B. Bausell, and J. R. Jenkins. "The Effect of Class Six on the Learning of Mathematics: A Parametric Study with Fourth Grade Students." *Journal for Research in Mathematics Education* 4 (May 1973):170–176.

A proposal for the improvement of the mathematics training of elementary school teachers

HERBERT F. SPITZER

The University of Texas, Austin, Texas

*Herbert Spitzer is, this year, visiting professor of
mathematics education. He is also assisting with the
Elementary Mathematics Migrant Program of the Southwest
Educational Development Laboratory.*

Leaders in mathematics education have long recognized that the elementary school teacher's lack of mathematical background was a major deterrent to improvement of instruction in this curricular area. While a number of recommendations for improving the mathematics part of teacher-education programs were made before the new mathematics era,[1] very little change in the mathematics preparation of elementary school teachers took place until after new mathematics began to be widely used in the elementary schools. Then, when teachers began to meet elements of strange new content, the need for additional mathematical knowledge became readily apparent.

The teacher-training panel of CUPM prepared a report published in 1961 recommending that the minimum mathematics training of elementary teachers include twelve semester hours of mathematics, with major emphasis on the arithmetic of the real numbers, introductory algebra, and informal geometry. College courses based on such recommendations, and books to be used in the courses, quickly came into existence. In spite of the fact that these courses and materials were prepared in response to a recognized need, they have apparently not been very successful in providing elementary school teachers with the mathematical knowledge needed to teach in many current elementary school programs. The apparent inadequacy of current college mathematics courses for elementary school is frequently mentioned by teachers of mathematics methods, by supervisors, and by the teachers who have taken the courses. Further indications of the inade-

---

[1] See Foster E. Grossnickle, "The Training of Teachers of Arithmetic," and C. V. Newsom, "Mathematical Background Needed by Teachers of Arithmetic," in *The Teaching of Arithmetic*, Fiftieth Yearbook, Part II, of the National Society for the Study of Education (Chicago: University of Chicago Press, 1951); and Arden K. Ruddell, Wilbur Dutton, and John Reckzeh, "Background Mathematics for Elementary Teachers," in *Instruction in Arithmetic*, Twenty-fifth Yearbook of the National Council of Teachers of Mathematics (Washington, D.C.: The Council, 1960), pp. 316–17.

quacy of the current mathematics programs for training elementary school teachers are found in such publications as "The 1967 Report of the Cambridge Conference on Teacher Training," "Improving Mathematics Education For Elementary School Teachers, A Report of a 1967 Conference at Michigan State," and the January, February, and March 1968 issues of THE ARITHMETIC TEACHER.

This dissatisfaction with the college mathematics courses for elementary teachers is a problem that warrants careful consideration. It seems to me that a major reason for students' dissatisfaction is their failure to see much relationship between what they study in these courses and their image of what mathematics they will teach to children in the elementary school.

Because the content for these mathematics courses was selected for its relevance to the K–6 elementary school mathematics program, it would seem that this criticism could hardly be valid. Yet it exists. Reasons for this are probably varied, but chief among them is the manner in which the content is presented—either by the college textbook or by the college instructor or by both.

To illustrate, consider a typical introduction to the study of geometry as presented in a representative mathematics book for elementary school teachers. First, there is a brief review of the concept of set, followed by a statement that study of sets of points is to be the next topic for consideration. Then this is followed by an allusion to the three infinite sets of points to be studied: the line, the plane, and all space. After representations of some subsets of planes and lines are presented, a statement is made that such representations are referred to as geometrical figures. Following four pages of such expository material, some study exercises are offered. These require the student to draw representations of a line, a plane, and a point; to use the line symbol and letters to name a line, and so on.

That the college freshman or sophomore fails to see a connection between such mathematical exposition and the curriculum of the kindergarten or first grade, where the college student hopes to teach in three or four years, is not at all surprising. The content presented in the college book is mathematically sound, and it is indeed relevant to kindergarten-primary mathematics. What, then, can be done to help college students recognize the relationship of this mathematics to their teaching, and what can be done to help them acquire an interest in the study of such content? One proven approach is illustrated by the following experience.

The study of geometry in a college class for elementary education majors began through consideration of the following dittoed material, describing a classroom activity.

KINDERGARTEN MATHEMATICS

In an elementary school, noted for its good mathematics program, a major kindergarten mathematics activity made use of three representations of intersecting geometric figures marked on the floor (Fig. 1). The side of the square was 3 feet long. Small wooden blocks in the shapes of squares, triangles, and circles were used in the activity. There were three colors (red, yellow, black), two thicknesses, and two area sizes of each block. At the beginning of the activity the blocks were randomly placed inside and (a few) outside the geometric figures on the floor.

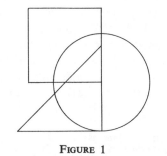

FIGURE 1

In this activity one child who is "It" selects a block mentally, tells the teacher, and then other children attempt to identify the selected block by asking "yes-or-no" questions (see dialogue below). A "yes" answer permits the questioner to continue. A "no" answer results in the right to ask questions being passed to another child. The child who identifies the block earns the privilege of being "It" for the next round of the activity. A report of the complete dialogue of one round of the activity in a class follows.

TEACHER: Janet has told me which block she has selected. Bill, you may ask questions first. Remember that your turn to ask questions stops when you get a "no" answer.

BILL: Is it in the round figure? (Yes.) Is it in the figure with four sides? (No.)

TEACHER: Your turn, Carl.

CARL: Is it a big block? (Yes.) Is it a thin block? (Yes.) Is it a red block? (No.)

TEACHER: Alice, it's your turn.

ALICE: Is it black? (No.)

TEACHER: Henry, it's your turn.

HENRY: Is it inside this space (*pointing to the space inside the triangle and circle*)? (No.)

TEACHER: Bill, it's your turn again.

BILL: Does it have sides? (Yes.) Is it in the three-sided figure? (No.)

TEACHER: Carl.

CARL: Is it yellow? (Yes.) Is it this one (*pointing to a large, thin, yellow, triangular block inside the circle, outside the square and the triangle*)? (Yes.)

TEACHER: Good, Carl. You are the winner.

If the answer had been "No," Janet would have been the winner and could have selected another block to start another round.

The teacher who directed this activity believes that pupils develop some important mathematical knowledge through the activity. What knowledge that could be useful to a child's mathematical growth might be developed in this way?

After a time for individual consideration of the question posed in the dittoed material, the students in the college class reacted. That at least some members of the class thought the idea portrayed in the material had merit is reflected in the following seven points recorded during the discussion:

1. Any location (in this activity, an object) is either inside or outside of a given closed curve.

2. Some figures have round edges; others have line segments for edges.

3. Numbers may be helpful in identifying figures.

4. Given information is most useful if a logical pattern of thinking is followed. For example, if the object is within the circular region, then parts of the square region and the triangular region are excluded.

5. There are different kinds of closed curves.

6. There is a need for accurate descriptions. For example: "Are the sides of the figure the same length?"

7. The longer accurate descriptions can be discarded when the shorter name (for example, "square") is recognized by all.

In the discussion of the above seven statements, and of other aspects of the kindergarten mathematics activity, some questions that led to further study were asked. Among them:

a) What is a closed curve? This question was formulated after a student, in referring to statement (1) above, asked, "Why do you say 'closed curve'? Wouldn't 'figure' be better?"

b) What is a curve? This question arose during the consideration of the preceding question.

c) Couldn't the location (see statement *a*) be on the closed curve?

d) Why use circular region, square region, and triangular region? (See point 4 above.)

e) What are the really distinguishing characteristics of each of the geometric figures used in this activity?

It is my firm belief that the study of geometry, guided by such questions as those given in the preceding sections, proves to be no less fruitful than study guided by questions and suggestions based on material in current college textbooks. Such an appoach to the study of geometry makes it easier for college students to recognize that college mathematics is indeed relevant to elementary school mathematics.

It also seems to me that the time has arrived when "elementary school teachers-to-be" should expect that the required college mathematics classes, and the materials of instruction used in those classes, reflect the kind of a mathematics program that is envisioned for the elementary schools where these students are to teach.

History of mathematics in the training program for elementary teachers

RICHARD L. FRANCIS

A professor of mathematics and education at Southeast Missouri State University, Richard Francis teaches undergraduate and graduate courses for elementary and secondary teachers.

The implications of an appreciation of mathematics on the part of prospective elementary teachers are far-reaching, considering the teaching responsibilities they are about to assume. In light of these far-reaching implications, the college mathematics program for elementary teachers presents a definite challenge. One promising and meaningful provision for developing mathematical appreciation falls into the area of the history of mathematics.

Virtually no one, prospective teacher or otherwise, could fail to name a great musician if asked (say Brahms or Bach or Beethoven). The case for mathematics is another matter. The odds for success in naming a few great mathematicians are less than encouraging. Such a comparison points to a real deficiency, which if corrected has tremendous potential for improving the teaching of mathematics, whatever the level. Our concern at this time is the mathematics offering for the student soon to embark on the profession of elementary teaching.

The Committee on the Undergraduate Program in Mathematics (CUPM) has set forth in its recommendations for all elementary school teachers the following:

1. The structure of the number system (two semesters)

2. Elementary contemporary algebra (one semester)

3. Informal geometry (one semester)

Each of these areas lends itself well to the inclusion of much mathematical history, a major purpose being that of motivation and appreciation. Two safeguards should however be kept in mind:

Balance. Vital topics, such as those just cited, cannot be compromised in the providing of such supplementary materials as mathematical history would imply.

Relevance. The appropriateness of the considered historical items is determined essentially by the vital topics and course objectives.

In the case for balance, topics such as the structure of the number system cannot be permitted to degenerate into little more than so much history. Relevance to the topics is hardly a severe restriction; a wealth of material is available, both topically and at the right level.

Prospective teachers may not eventually use in actual teaching everything of a historical nature presented to them. However the inclination towards the use of historical references in teaching is a desired trait in teacher resourcefulness, making the provision entirely worthwhile.

Courses as ordinarily taught (number systems, algebra, and informal geometry) are not completely devoid of historical sidelights. Such overtones are evident in many of the stressed topics. Consider the following illustrations, using familiar items in the elementary teacher sequence:

1. *Important labels* (Hindu-Arabic numerals, Cartesian coordinates, Venn diagrams, etc.)
2. *Significant statements* (theorem of Pythagoras, Playfair's axiom, etc.)
3. *Various properties* (Archimedean property, Dedekind property, etc.)
4. *Certain methods* (Euclidean algorithm, sieve of Eratosthenes, etc.)
5. *Formulas* (Euler's polyhedral formula, De Morgan's formulas, etc.)
6. *Symbols* (origins of basic operation and relation symbols such as $+$, $-$, $\times$, $=$, etc.)
7. *Vocabulary* (origins and literal meanings of such words as geometry, locus, parabola, etc.)

A rich history underlies concepts such as these, revealing the contributions of diverse groups of people as well as that of specific persons. The student naturally wonders why certain names are attached to so many of the concepts studied. This is an ideal opportunity for capitalizing on the learner's curiosity. The teaching technique of "getting on with the problems" often rules out any effort in this direction.

The historical emphasis need not be restricted to names of concepts however; these references are fairly pointed and specific. Mathematical history also has applications in a broader, general manner. Consider, for example, the fact that the child's formal encounter with number classes parallels perfectly mankind's historical development of number types. As number set extensions are made in the classroom, many of the problems associated with the historical extensions bear mentioning. These range from the inadequacy of counting numbers in an increasingly technical society to the motivation underlying many number type

names. Certain name choices, such as *irrational* or *imaginary,* are very meaningful when analyzed. An instructive activity at this point involves consideration of why misnomers persist in present day mathematical vocabulary, and whether or not similar problems exist in other disciplines.

Instances from the past wherein mathematicians have anticipated the needs of modern society also prove enlightening. Though practicality and application are not the sole concern in justifying the topics studied, students are apt to reason this way. A recurring question in mathematics teaching is Why study this topic? or What is this topic good for? References to projected usefulness assist at this point. Such references are reinforced by illustrations of past mathematicians having anticipated the needs of a later society. These may range from the development of spherical geometry at a time when the earth was thought to be flat, to the last century explorations of nondecimal number bases, well in advance of the modern day computer.

Topics could be timely as well. In light of the imminent conversion to metric measure and the frequent advice to "think metric," some consideration could be given to the historical background of the metric system. A description of the times that brought the metric system into being would make the present transition all the more meaningful. Problems from such a time period (the late eighteenth century) have a familiar ring ranging from poorly chosen units of measure to an intense need for standardization. Of special historical significance is the original intention as to what this fundamental unit of measure should be (i.e., one tenmillionth of the earth's quadrant, say from the North Pole to the equator). Subsequent efforts to obtain this length, along with the problems of properly preserving the established measure, make any discussion of the metric system all the more interesting.

Many historical references are justifiable simply as "attention grabbers." What student doesn't sit up and listen when one discusses, say, legislating the value of pi (House Bill 246, Indiana Legislature, 1897) or subtle Biblical references to pi (1 Kings 7:23 or 2 Chronicles 4:2). Historical attempts to better approximate the number make interesting reading as well. Or perhaps in discussing numeration, the classical number names could be augmented by a comment · concerning Edward Kasner's coining of the word *googol,* representing 10^{100} (along with the more fascinating *googolplex*). These terms invariably stimulate interest.

All of the foregoing are simply illustrative. There are many other possibilities for the use of historical references. Moreover, activities of various kinds—compiling a collection of pictures of well-known mathematicians or preparing time lines, which give a chronological perspective—supplement discussions such as those suggested.

A proper use of auxiliary and supplementary materials is never an easy task, but this is part of the challenge to providing better for the needs of prospective teachers. The courses that are taught may well be the person's final encounter with mathematics in a formal learning situation. Many negative attitudes still persist, even to this late date in the training program. Too many have the feeling that "mathematics is basically uninteresting and less than exciting to teach." The courses that prospective teachers take may be a last opportunity for acquiring an appreciation of mathematics. In light of the concern that elementary teachers be good and enthusiastic mathematics teachers, mathematics history is a promising device, a tool so easily overlooked and too often uanappreciated.

References

Bell, E. T. *Men of Mathematics.* New York: Simon & Schuster, 1937.

Cajori, Florian. *A History of Elementary Mathematics.* New York: Macmillan Co., 1950.

Eves, Howard. *An Introduction to the History of Mathematics.* New York: Holt, Rinehart and Winston, 1953.

————. *The Other Side of the Equation.* Boston: Prindle, Weber & Schmidt, 1971.

National Council of Teachers of Mathematics. *Historical Topics for the Mathematics Classroom.* Thirty-first Yearbook of the National Council of Teachers of Mathematics. Washington, D.C.: The Council, 1969.

Read, Cecil B. "The History of Mathematics—A Bibliography of Articles in English Appearing in Seven Periodicals." *School Science and Mathematics* 66 (February 1966).

Some Reflections on Teaching Mathematics for Elementary School Teachers

By Lowell Leake

Virtually every college or university training elementary school teachers offers a course in mathematics content with a title similar to "Mathematics for the Elementary School Teacher", the title we use at the University of Cincinnati. The course is usually a year long and in most states it is a certification requirement for grades kindergarten through eight; it evolved from the recommendations of the Committee on the Undergraduate Program in Mathematics (CUPM) of the Mathematical Association of America in the early 1960s. Little, however, has appeared in journals such as the *Arithmetic Teacher* about the experiences various universities or instructors have had with this course. This paper describes some aspects of such a course at the University of Cincinnati, and I have had two major purposes in mind in writing it. One is to communicate some successful ideas to others who teach similar courses, and the other is to encourage others to submit their ideas, experiences, and suggestions so that all of us can share and profit from each other's successes and failures. A third possibility is to generate some comments from teachers in elementary and middle schools who once took such a course and now, in retrospect, have suggestions to make to those of us who teach it.

Lowell Leake is a professor in the Department of Mathematical Sciences at the University of Cincinnati in Cincinnati, Ohio. Prior to his university teaching, he taught in the junior and senior high school. He was one of the translators of The Origin of the Idea of Chance in Children *by Piaget and Inhelder and recently served a three-year term as a review editor for the* Mathematics Teacher.

One of the most successful ideas I have used is to require that my students read articles in the *Arithmetic Teacher* related to the content we are studying that quarter. In each of the three quarters of my section of our course, students who want to receive an A must, in addition to meeting the standards required on tests, and so on, read three articles from recent issues of the *Arithmetic Teacher*. For each article I require a one- or two-page "reaction paper" to be handed in any time prior to the last day of class for that quarter. I ask students to include both a short summary of the article and their own personal reaction to it in terms of their perception of the article's quality and the reasons for their assessment. I do not grade the papers but I read all of them carefully. To get a B two such reports must be handed in, and C and D students must hand in one paper.

For several years I superimposed on this requirement the option of reading up to five articles each quarter (with reaction papers) for anyone, and I added one point per paper to the overall numerical average of the student. I abandoned this approach as too generous since it inflated a number of grades that I felt should not have been raised.

The students' response to this reading requirement, once they get over the initial shock on hearing about it, has been totally positive. Not only do their comments in the reaction papers indicate curiosity and enthusiasm, but they also indicate clearly that the readings help them relate the course content to their future careers in the classroom. It has done a great deal to relate content to methodology without taking any time from classroom instruction. The other clear benefit is that when these students graduate they have a personal acquaintance with a journal that may turn out to be their best friend in the classroom.

In Cincinnati we have an active group, the Mathematics Club of Greater Cincinnati, which holds frequent meetings. We are also, from time to time, the host city for an NCTM convention. When these meetings occur I encourage my students to attend, and I allow them to substitute a report on the attendance at a section meeting for elementary teachers for one of the reaction papers. A few students have done this, but not very many.

An alternative to assigning readings in the *Arithmetic Teacher* itself is to require that the students purchase a good book of readings, which would of course include articles from many other journals. An excellent choice for this would be a book such as *Activity Oriented Mathematics*, edited by Schall (1976). The advantage of this approach is that the students become exposed to the idea that many good journals exist that can help the elementary classroom teacher become a better mathematics teacher. The disadvantages are the expense to the student and the fact that students do not actually see any particular journal and what it has to offer on an issue-by-issue basis.

Another aspect of my experience that has been quite successful, but for relatively few students, has been an arrangement to match bright, able students on a one-to-one tutoring basis with students who are struggling to pass. This is strictly an optional arrangement. After the first hour-long examination, students who received an A may volunteer to be a tutor for any students who want help. Both those who want to tutor and those who want to be tutored let me know about this in writing and I try to match them up. Students who tutor are excused from the final examination, but they must take the other hour-long examinations (a total of three each quarter) and they must do the readings described earlier. They also are required to attend class, to meet with the student or students they tutor at least twice a week, and to turn in a written summary (3–5 pages) of the tutoring experience. I explain at the outset that this is a very difficult way to earn an A; it would be far easier to take the final. On the other hand, I point out that they will learn much more if they have to teach the material, both about the content and about the learning problems others face, and that they will gain a good deal of satisfaction from helping fellow students. On the average, about 10 percent of the A students volunteer for this experience. Sometimes they have to tutor two students (if they agree) since I can't guarantee a one-to-one correspondence between those helping and those needing help. The students who get the help have the advantage of free tutoring from someone who actually sits in on the lectures and takes the tests. I am toying with the idea of requiring this tutoring experience for anyone who wants an A, but I am not sure this would be a wise step.

A third extremely successful experiment of the things I have tried has been to have each student buy a calculator. Most buy inexpensive but adequate four-function calculators. Some have better machines, with keys for square roots, squares, reciprocals, and so on, and some have very good scientific calculators. Before they buy—only about 25 percent do not already own one—I suggest what they should look for.

The students use the calculators for homework, classwork, and on most tests. When I first announce this requirement a few students grumble, but these expressions of dissatisfaction evaporate almost immediately. To satisfy my own curiosity I always ask the class, when I first announce this requirement, how many of them disapprove the use of calculators in elementary school. The vast majority do not approve, but at the end of the year, when I repeat the same question, almost no one disapproves. Of course, during the year I make sure that instruction on when to use calculators, how to use them, and the pros and cons of their use are an integral part of the course. As most readers realize, calculators not only help with complicated calculations, but they also allow the use of more realistic data in problems. Furthermore, they definitely help to teach concepts. For example, in studying square roots, my students are asked to find square roots without using the square root key—by guessing and adjusting the guess. The same idea works for cube roots. If you want to discuss this method more formally, using what is known as Newton's method, the calculator makes it possible to do the iterations for the third and fourth approximations (and beyond), whereas paper-and-pencil techniques bog down at the third approximation because of the messy arithmetic. Another example is in studying the distinction between repeating and terminating decimals; calculators are a great help with this idea. Most important, perhaps, calculators allow students to concentrate on what to do in solving a mathematics problem, rather than on the details of time-consuming arithmetic.

My classes have been using calculators since 1975–76, and I feel so strongly about this idea that I believe anyone who writes a text for the content course in elementary school mathematics should include a chapter or units on the calculator and have some exercises in almost every section keyed for calculator solutions. And anyone who teaches this course without calculators is missing the best opportunity there is to provide the kind of instruction relating to calculators that every preservice teacher needs. The calculator should be part of the instructional strategy for every instructor teaching this course.

Another idea we have tried and still use at Cincinnati is our eligibility test (20 simple items) for students to pass at the beginning of the year on basic skills with whole numbers, common fractions, decimals, and percents. If they do not pass this test (at least 14 correct), which is given on the second day of class, after announcing it the first day, they must drop the course and enter a special course in remedial arithmetic. This latter course is taught on an individualized basis and students may pass it quickly if they work hard. Then they can re-enter the main course in a trailer section during the winter quarter.

Most of these remedial students, about 90 percent over the past five years, have eventually become certified, and that has caused me some genuine concern. Is our remedial course so good that the students suddenly blossom into adequate mathematics students? Or, is our main course too easy and are we letting very marginal students become teachers? This has led me to think that we ought to have an exit test on basic skills (minimum compentencies?) for all students before they receive their grade for the final quarter of our course. What do other universities do? It also makes me think that even if our course covers the content in a method that stresses intuition rather than rigor, we ought to demand a high degree of mastery in this course for a passing grade.

Another thought is this. We have seen a significant decline in enrollments in elementary education over the past few years, and I assume this is a common experience for most schools. I have a sinking feeling that the drop in enrollments has largely been among the most able students—for two reasons. One, the more able students are more alert to the economics of the labor market and know that teaching may be a poor choice of careers in terms of opportunities and economic gain. Second, these students, being bright, have much more flexibility and opportunity concerning the choice of alternate careers. For example, opportunities are good in business administration and this is particularly true at present for women. Are we, then, not only training fewer teachers but are we possibly training proportionally more marginal teachers than in the past? Does anyone have any hard data on this subject? It certainly seems to be a question that lends itself to statistical analysis.

Two other ideas that have proved successful and popular with students relate to the teaching of units on probability and statistics, and geometry. In probability, although I cover whatever the text has to offer, I concentrate on a Monte Carlo approach for approximating probabilities. We start with dice, cards, and coins but eventually we turn to random-digit tables. This seems to be far more successful and exciting than a more formal approach stressing formulas for combinations and permutations (although I discuss these, too, after Monte Carlo techniques). Piaget and Inhelder (1975) have told us that true understanding of probability comes after the acquisition of formal, logical operations. Some of our preservice teachers have not, unfortunately, reached that stage of cognitive development. Therefore, it makes sense to keep our instruction in probability tied to concrete operational techniques, and the Monte Carlo approach certainly does this. Besides, and perhaps just as important, a Monte Carlo approach gives our students specific experience on methodology in the content course on how to teach probability in the upper grades. This gives us an unusual chance to teach methodology by example imbedded in the content course being covered.

In geometry, I feel the study of isometric transformations should be a major part of the material covered. This can be done very successfully with an activity approach via tracing techniques. The ideas are beautifully described by Phillips and Zwoyer (1969) in *Motion Geometry*. The students are fascinated by this approach and, once again, they learn some difficult material in a concrete operations setting, which they can later duplicate in their own classrooms. Perhaps the most important aspect of this approach is that it uses familiar materials from tenth-grade geometry to develop an entirely new attitude towards geometry. Too often the material on geometry is either too dull or too repetitious of tenth-grade geometry. Use of tracing techniques and isometric transformations avoids both of these pitfalls.

For the past few years I have distributed copies of the basic mathematical skills developed by the National Council of Supervisors of Mathematics (see *Arithmetic Teacher*, October 1977) to each of my students. I lecture on this statement and attempt to show by example throughout the course that computation is only one of ten basic skills (and one that can be simplified by the calculator). The students know that a question about this statement will appear on at least one test.

Max Bell, in his address at the 55th Annual Meeting of the NCTM in Cincinnati in April 1977, cogently pointed out that calculation is but one of the ten objectives on the NCSM list. He observed that if we stress only that objective of calculation, and the public decides a few years from now that hand-held calculators can take care of that part of the curriculum, mathematics training in the elementary school may be subject to enormous pressures to reduce its time demands on our students. Then the other nine objectives may be thrown out with calculation, an eventuality that would have long-lasting and tragic consequences. In April 1980 Bell reemphasized this point in his address at the 58th Annual Meeting of the NCTM in Seattle when he spoke on NCTM's Recommendations for School Mathematics of the 1980s.

We must convince everyone that the other nine objectives are just as much mathematics, indeed more, than the ability to do calculations. The place to start, of course, is to convince preservice elementary school teachers of the existence and importance of all ten objectives. Proper use of the calculator in the preservice content course will help teachers understand the necessity of teaching arithmetic with all ten objectives in mind.

Near the end of the school year, I devote one lecture to giving a sales pitch for joining NCTM. I describe its activities, pass out NCTM information brochures, publication lists, and copies of the *Arithmetic Teacher*, and stress the student membership rate. Then I hope for the best. A very gratifying number of students have joined the NCTM this way. Personally, I feel that the experience of reading articles from the *Arithmetic Teacher* throughout the year helps to convince these students that membership in NCTM is definitely worth the price.

Finally, I am trying to add another dimension to our course, problem solving. It seems to me that if we want to emphasize problem solving in the 1980s, the place to start is in the training of preservice elementary school teachers, and this is something we can accomplish almost immediately. The basic structure of "Mathematics for the Elementary School Teacher" need not change to get things rolling. A problem-solving atmosphere can be injected into existing courses using standard textbooks, if an instructor is willing to do a little extra work and use a little ingenuity. To get started, an instructor might devote 15 minutes of each class meeting to problem solving, or one day a week. Perhaps a laboratory session can fit into the schedule for this purpose. The latest NCTM Yearbook, *Problem Solving in School Mathematics*, is filled with excellent advice and ideas that can be incorporated into the content course. Perhaps that yearbook ought to be required as a supplemental text for all students in the content course. Regular assignments and lectures concerning its content could turn the course into something that will be much more effective for our future teachers.

There are only a few ideas, but certainly enough for this paper. I hope many readers will decide to try one or two of them, but I hope even more others will submit similar papers to the *Arithmetic Teacher* to describe other ideas that could benefit all of us.

References

"National Council of Supervisors of Mathematics Position Paper on Basic Skills." *Arithmetic Teacher* 25 (October 1977): 19–22.

National Council of Teachers of Mathematics. *Problem Solving in School Mathematics.* 1980 Yearbook. Reston, Va.: The Council, 1980.

Phillips, J. McKeeby, and Russell E. Zwoyer. *Motion Geometry.* New York: Harper and Row, Inc., 1969.

Piaget, Jean, and Barbel Inhelder. *The Origin of the Idea of Chance in Children.* New York: W. W. Norton and Company, Inc., 1975.

Schall, William E. *Activity Oriented Mathematics.* Boston, Mass.: Prindle, Weber, and Schmidt, 1976. ♥

More Reflections on Teaching Mathematics for Elementary School Teachers

By **Rick Billstein** *and* **Johnny W. Lott**

In his article, "Some Reflections on Teaching Mathematics for Elementary School Teachers," (November 1980 issue of the *Arithmetic Teacher*), Lowell Leake suggested that mathematics educators share teaching ideas, successes, and failures in courses for elementary school teachers. This article is a response to his suggestion.

For many years now at the University of Montana, we have had a sequence of three one-quarter, mathematics content courses for elementary school teachers. The courses, which developed along the lines of the recommendations of the Committee on the Undergraduate Program in Mathematics (CUPM), now include many of the facets described by Leake, but are handled in a somewhat different manner.

In their methods classes, which are separate from the content courses, our students are required to start a card file of activities developed through the use of articles in the *Arithmetic Teacher*. In our opinion, the card file is more important, in the long run, than reaction papers.

For the content courses, we have found that by supplying the students

Rick Billstein in an associate professor of mathematics and associate chairman of the mathematics department at the University of Montana in Missoula. His principal responsibilities are in the area of elementary school mathematics education. Johnny Lott is an associate professor of mathematics on the same campus. His specific responsibilities are in the area of mathematics education for elementary and secondary school teachers. He has taught in the secondary schools and during the current academic year he is on sabbatical leave from the university to teach in the Pelican, Alaska, Schools.

with lists of questions that could have come from the classroom, similar to those posed by Crouse and Sloyer (1977), and a bibliography with references to the AT and other sources, we can require that the students write responses to the specific mathematical questions by doing research in both the content and method areas. These questions and bibliographies are chosen in conjunction with the topic being discussed and are continually updated.

The following is an example of a question we might ask when integers are the subject of study:

> A student says that $^-5 > {}^-2$, since a debt of \$5 is larger than a debt of \$2. What error is being made by the students and how can it be corrected?

We think that by using the AT and other library resources our students will realize that they can find answers and help for classroom questions. To further emphasize the usefulness of these resources, we strongly encourage our students to take advantage of their student status to become members (at reduced rates) of both the NCTM and the Montana Council of Teachers of Mathematics (MCTM).

Occasionally we invite a practicing elementary school teacher to teach our classes and to answer questions for our students. This practice, along with that of bringing various textbook series to class, serves to give us more credibility concerning much of the material being covered. Our students are also asked to review at least one textbook in the lower grades (K through 4) and one in the upper grades

(5 through 8) each quarter. In this way, they can see just how the mathematics we are covering in our classes is presented at those levels.

Calculators have been an integral part of our courses since the MCTM conducted its Calculators in Schools Project in 1977. Since most of our students come to us with little knowledge of calculating devices, we start with the very basics—how calculators can be used in the classroom, which types of calculators are available, the type of logic the calculator should have, the uses of various keys, and the things to be aware of when ordering a classroom set of calculators.

Forty calculators are available for use with our classes and we encourage students to buy their own machines only after we have discussed the pros and cons of various calculators. Once the preliminaries are completed, we give our students a set of selected problems for various grade levels both to work and to discuss in class. The current research on the use of calculators in the elementary school is also investigated. Microcomputer and computer demonstrations are also arranged as time permits. An eventual goal is to integrate their use more fully into our courses.

Two years ago we made problem solving the focus of our courses. The first topic of the first quarter is a four-step, Polya-type approach to solving problems. The following four steps are used: understanding the problem, devising a plan, carrying out the plan, and looking back. Initially, the problems investigated draw on few mathematical skills, but do require creative thinking. The various problem-solving strategies, such as looking at patterns and reducing the problem to a simpler problem, are discussed and emphasized. The following is a typical example:

It is the first day of a mathematics class and twenty people are present in the room. To become acquainted, each person shakes hands just once with everyone else. How many handshakes take place?

In this example, the strategies discussed are trying a simpler problem, making a table, looking for a pattern, and drawing a model.

After the initial presentation of problem solving, we introduce each new unit with a somewhat sophisticated preliminary problem, stated in simple terms, which can be solved using problem-solving techniques along with the mathematical content of the unit to be taught. Students are encouraged to try to solve the problem immediately, but very seldom is this done. As the students work through the block of material in the unit, they are to keep the preliminary problem in mind and whenever they think they have learned enough content to solve the problem, they are to try it again. If the problem is not solved by the time the unit is completed, then the problem is discussed and solved in class.

The following is an example of a preliminary problem for a geometry unit:

Suppose a stained-glass-window maker needed assorted triangular pieces of glass for a project. One rectangular plate of glass that she planned to cut contained ten air bubbles, no three of which were in a straight line. To avoid having an air bubble in her finished product, she decided to cut triangular pieces by making the air bubbles and the corners of the plate the vertices of the triangle. How many triangular pieces did she cut?

In the teaching of probability, we have found that hands-on experience with devices that generate random events is important and that teaching probability using a formula-oriented approach is not as effective as developing almost all the material via tree diagrams. The Monte Carlo approach is very helpful in developing many of the ideas in the course. Even when students find that they cannot work many problems theoretically, they find ways to simulate the problem using random-digit tables. Simulation techniques are very powerful and can be reinforced even more if computer simulation techniques are added to the course.

For the geometry units, we have chosen a construction approach. By having students involved in paper-folding activities, and compass and straightedge and MIRA constructions, most plane geometry notions can be motivated. We find the MIRA is a very valuable tool in developing geometrical concepts (see Woodward 1977). All compass and straightedge constructions, except drawing a circle, can be investigated using a MIRA. MIRAs are also especially helpful when studying line symmetries and motion geometry.

A discussion and a showing of drawings by M. C. Escher are valuable motivators in the study of isometric transformations. Many drawings are analyzed and transformations are used by students to develop Escher-type drawings of their own.

In our classes, the study of metric measure is integrated into the course. It is introduced in a hands-on setting and is not taught in terms of conversions from the customary to the metric system. Conversions from commonly used metric units to other commonly used metric units are included. Students are also made aware of the relationship between the metric units of length, volume, capacity, and mass (weight). A discussion of the history of the metric system, why we are going metric, and the current status of the metric changeover is also included.

One of the most important things that teachers of courses for elementary school teachers can do is to encourage their students to attend at least one NCTM meeting per year. We not only encourage students to attend, but also recommend specific sessions and speakers for their first meetings. Meetings can give prospective teachers an encouragement far beyond normal classroom experiences. We find that as college teachers we also benefit from NCTM meetings and publications. Through these experiences we can continue to keep current in our field and can continually revamp our courses based on new information.

These are but a few of the ideas which have worked well for us in our teaching. We hope we have fulfilled some of Leake's expectations and that this article will serve to motivate even more discussion.

References

Billstein, Rick, Shlomo Libeskind, and Johnny W. Lott. *A Problem-Solving Approach to Mathematics for Elementary Teachers.* Menlo Park: Benjamin Cummings Company, 1981.

Crouse, Richard J., and Clifford W. Sloyer. *Mathematical Questions from the Classroom.* Boston: Prindle, Weber, and Schmidt, 1977.

Wilson, Art. *The Effects of the Hand-held Calculator upon Achievement Test Scores of Elementary School Mathematics Students.* Unpublished dissertation at University of Montana, 1978.

Woodward, Ernest. "Geometry with a Mira." *Arithmetic Teacher* 24 (February 1977):117–18. ◆

CONTENT/METHODS COURSES

Training elementary mathematics teachers in a one-semester course

WILLIAM B. CRITTENDEN

An associate professor in education at Houston Baptist
University, William Crittenden is the only member of the faculty
teaching mathematics education. He has sought a model that would yield
the best possible results in training elementary teachers in
one semester. He has also been a leader in the Houston
area in the· evolution of performance-based training.

When the writer joined the faculty of Houston Baptist University in 1968, he was given the assignment of prescribing the mathematics education training experiences for prospective elementary teachers. Since the college is primarily for the liberal arts and the curriculum includes a large number of required cources, there was only one three-semester-hour course available in which elementary teachers could be prepared in mathematics. A "prepared" teacher was defined as one who had—

- achieved a prescribed level of mastery of mathematics content;
- accumulated a theoretical and empirical repertoire of teaching strategies, techniques, aids, and activities; and
- exhibited a positive attitude toward mathematics as a field of study.

The challenge was obvious and almost overwhelming. How could all three vital objectives be achieved in 45 hours of classwork during a single semester while the students were involved in from four to six other courses?

No other mathematics courses were required for elementary education majors at Houston Baptist University. The variability in requirements from college to college in America in this field was highlighted by Fisher (1967) who found that 40% of 78 teacher-training institutions graduated elementary teachers with three semester hours' credit (in mathematics or mathematics education) or less, and 90% required six hours or less.

The problem at Houston Baptist University became acutely clear as early attempts to solve it resulted in failures. If content were emphasized and methods ignored, students later complained (justifiably) that they had received little or no preparation in how to teach mathematics. On the other hand, when methods were emphasized and content ignored, supervisors of student teaching complained that the students so prepared made embarrassing content errors in their teaching efforts. When scales were used on a pretest-posttest basis to measure positive gains in attitude toward mathematics, it was found that few students showed changes of attitude and, when they did change, the change was just as apt to be negative as positive.

It was decided that more time for the training was absolutely essential. Even though it was not possible to award more than three hours' credit, a laboratory was scheduled for a second hour three times per week. Thus, the class time was increased from 45 clock hours per semester to 90. Simultaneously, arrangements were made with a nearby public elementary school to permit classroom teaching experiences collaterally with campus seminars.

Since this methods course occupied a position in the training sequence between the classroom observations of the sophomore education major and the student teaching of the senior education major, a descriptive name for this intermediate activity was needed. The program was referred to as "student preteaching in elementary mathematics," and the participants became known as "student preteachers."

Students in the program were given pretests in mathematics content, attitudes toward mathematics, and knowledge of mathematics teaching methods. Those who failed to achieve 90% level of mastery of content were given individually prescribed programs in the areas of greatest weakness. No formal lectures in mathematics content were delivered during the semester, but free time was allowed during seminars to provide opportunity for learners to ask for needed explanations. All inquiries were satisfied. Checks of students' work on the programs were made as a control measure.

Pretest results revealed that 95% of the students were deficient in their knowledge of elementary school mathematics; more than 50% of them disliked or feared mathematics and lacked confidence in their ability to teach it; and all of them had a limited knowledge of mathematics teaching methods. Most of them reported that their referent source for teaching techniques was to be found in the way their elementary school teachers had taught them.

With excellent cooperation from the laboratory school staff, the student preteachers were assigned to classes that had been divided into small groups on the basis of mathematics ability. The host school had reorganized schedules to permit the intraclassroom group teaching experiences to occur simultaneously, thus permitting the preteachers to form carpools. The laboratory teaching proceeded two days per week with a two-hour seminar on the college campus one day each week.

The on-campus seminar was devoted to building and demonstrating teaching aids. The elementary school mathematics curriculum was summarized in ten broad concept areas, and a team report was prepared for each concept area. Games, activities, developmental techniques, manipulative and demonstrative aids, and evaluative methods were surveyed. Each team produced a 30-page handout booklet containing drawings and descriptions of the teaching aids associated with each concept area. At the end of the course, each student had a collection of ideas for aids and techniques in teaching elementary mathematics that could be useful as a reference source.

The student preteachers were given teaching responsibilities immediately, each one teaching a group of pupils from the beginning day of his field experience. This precipitant beginning was ill-advised. The cooperating teachers complained that the preteachers made arithmetic errors in their teaching, and they resented being made responsible for teaching mathematics content to the preteachers. In addition, the cooperating teachers expected that the preteachers would use huge quantities of teaching aids, which they didn't; preteachers tended to rely predominantly on workbooks and duplicated worksheets, especially in the beginning of the course.

The student preteachers were panic-stricken at being given such heavy responsibilities so quickly when they were well aware of their state of unpreparedness. Since most of them feared or disliked mathematics anyway, the required teaching of it in the presence of experienced teachers, principals, or college observers, with no preliminary preparation acted as a crystallizing agent on their attitudes, rather than as a stimulus to positive change.

Through trial and error, the student preteachers *did* learn to adjust their teaching methods to the levels of their pupils. They *did* learn to manufacture and use teaching aids. And they *did* learn more mathematics content as a result of the seminar and laboratory experiences. Posttests showed significant gains in knowledge of mathematics content and teaching methods. However, the attitudes of the preteachers toward mathematics regressed, being more negative at the completion of the course than at the beginning.

The effect of the program upon the pupils in the host school was indeterminate. Subjectively, everyone involved in the program agreed that pupils responded positively to the use of aids and games, but no evidence concerning their achievement gains or losses was available. The cooperating teachers complained that there was a lack of continuity in the program, caused by the presence of preteachers only two days per week. They were not eager to continue the program in other semesters, even with revisions calculated to eliminate mistakes.

Even though the student preteaching program as described achieved two of the three essential objectives, it was declared a failure. To be considered "prepared," an elementary teacher must have knowledge of mathematics content, mathematics teaching methods, *and* a wholesome attitude toward the subject. The latter objective was still unachieved.

Determined to press forward in the search for a model which offered promise in achieving all three objectives, the writer and his colleagues in the Education Division analyzed the previous efforts and submitted a new proposal which they believed would eliminate the earlier flaws. It was decided to keep all factors in the training model constant except the entry to the full responsibility of teaching. What had proved to be a frightening experience because of the abruptness of entering the teaching process with no preparation might be turned into a constructive, gratifying experience by first preparing the preteacher. Thus, a phase-in process in four steps, with each new step requiring a higher level of responsibility, culminating in full teaching during the last part of the program was designed. By so delaying the ultimate burden of teaching until the students felt themselves to be better prepared, it was believed positive attitude changes could be effected.

The Research Council of the University approved a grant of funds for the new project in the spring semester of 1971. The new proposal called for the controlled phase-in of the student preteachers to the actual teaching responsibility, and it provided for an evaluation of the program effects on the pupils of the host classrooms.

Another school (Lakeview Elementary School, Fort Bend Independent School District, Sugarland, Texas) was enlisted to serve as a laboratory for the project. Careful orientation to the new program was provided the faculty of the new school, and volunteers were selected to serve as cooperating teachers. In addition, an equal number of control teachers were selected to provide data for comparisons of pupil achievement and attitude gains.

The actual experiences provided for the student preteachers were approximately the same as described in the earlier experiment, except for the phase-in sequence. The 16-week semester was divided into four phases of four weeks' duration each. (See fig. 1.) Phase I was the *Preparation* phase, and the group remained on the college campus meeting in seminars six hours weekly. Intensive work on mathematics content and teaching methods was accomplished during this period. The students were trained in construction of teaching aids. Practice in intraclassroom grouping and achievement of the "restaurant effect" (Creswell and Crittenden 1971) were undertaken. A conscious attempt to "shore up" the confidence of the preteachers was initiated.

Phase II of the project, labeled *Observation*, involved the assignment of preteachers to classrooms where they observed the cooperating teachers and pupils during the mathematics classes. The preteachers were

First training model (not recommended)

Student preteaching two days weekly for entire semester Seminar on content and methods one day weekly																
0	1	2	3	4	5	6	7	8	9	10	11	12	13	14	15	16

Weeks

Second training model (recommended)

Phase I Preparation	Phase II Observation	Phase III Tutoring	Phase IV Preteaching
Campus Seminars 6 hours weekly	Laboratory school observations Weekly	Laboratory school tutoring and teacher-aid activities 2 days weekly	Laboratory school full teaching responsibility 2 days weekly
Content and methods	Content and methods seminar 1 day weekly	Continuation of seminar 1 day weekly	Continuation of seminar 1 day weekly

0	1	2	3	4	5	6	7	8	9	10	11	12	13	14	15	16

Weeks

Fig. 1

trained in the use of office machines, audiovisual equipment, and various teaching aids. Their seminars on the college campus were devoted to a continuation of skill development.

Phase III, the *Tutoring* period, called for four weeks of tutoring pupils on a one-to-one, one-to-two, or one-to-three basis. By this time preteachers had been assigned their groups and they began to become intimately acquainted with the pupils they would eventually teach. Also, the preteachers had molecules-of-learning packets (Crittenden 1971) ready for tryout. During Phase IV, the *Preteaching* phase, the students had full responsibility for teaching their groups for two days per week for four weeks.

Results of the pretest-posttest data revealed that all three objectives for the course were attained. Students gained significantly in their knowledge of mathematics and mathematics teaching methods, and in addition a majority of them moved from negative to positive attitudes toward mathematics.

The actual classroom experience with grade school pupils is essential to the positive attitude change, but to assign the subject students the full teaching responsibility before they have developed self-confidence risks incurring the opposite result.

The ultimate dependent variables of grade-school pupils' (1) attitudes toward mathematics and (2) achievement while undergoing teaching by student preteaching teams were observed by comparisons between control and experimental samples. The Lakeview project involved seven sections of two matched samples of pupils in grades 1 through 5. There were two first-grade and two second-grade sections, as well as one each of third, fourth, and fifth grades in each sample. Matching was on the basis of previous mathematics achievement and IQ scores. Control teachers were aware of their participation in the project, but were told to teach their classes in a conventional manner.

Pretests of achievement and attitude were administered in February 1971. The California Achievement Test (Mathematics), 1970 Edition, Form A, Levels I through III, was used to measure achievement, and the Aiken–Dreger Revised Mathematics Attitude Scale (with an additional revision for use with grade school pupils) was employed to determine affective characteristics. Posttests using equivalent forms of the test instruments were given in May, and the resulting gains were compared by use of *t* tests.

Control pupils gained approximately six months in mathematics achievement during the four-month interval, and experimental pupils gained five months, a difference that was not statistically significant. It was concluded that the use of student preteachers as described in this paper did not affect pupil gains adversely or positively.

Pupil attitude changes during the treatment interval were also compared. The seventy-seven control pupils in grades 3 through 5 regressed slightly toward negative attitudes, while experimental pupils showed very little overall change of attitude. The difference was not statistically significant. When individual sections were compared, it was noted that the fourth-grade experimental pupils made significantly positive gains over the control pupils at the same level. On the basis of these findings, it was concluded that attitude changes due to using student preteachers did not occur. (Further information on data and statistical test results may be obtained from the author.)

In summary, the Lakeview project seems to offer a model for training elementary teachers in mathematics when the training institution is limited in the number of courses available. The program described in this paper showed that the objectives of preparing the prospective teachers in mathematics content, teaching methods, and positive attitudes are all possible of achievement.

The effects of the program on the host school were not so clear. The evidence indicates that there was no harm to grade school pupils as a result of the program, but no gains were effected either. However, the cooperating teachers enthusiastically endorsed the program and are eager to continue participating in it.

References

Creswell, John L., and William B. Crittenden. "Pre-Student Teaching in Elementary Mathematics: Theory Tested in the Classroom." *The Journal of Teacher Education* XXIII (Summer 1972): 211–214.

Creswell, John L., William B. Crittenden, and David Fitzgerald. "Conserving Teacher-Time: Effective Use of Intra-classroom Grouping Using Molecules-of-Learning Packets." *School Science and Mathematics* (December 1971): 769–74.

Fisher, John J. "Extent of Implementation of CUPM Level 1 Recommendations." ARITHMETIC TEACHER 14 (March 1967): 194–97.

Robinson, Jerry W., Jr., and William B. Crittenden. "Learning Modules: A Concept for Extension Educators?" *Journal of Extension* (Winter 1972): 35–44.

Preparing teachers for individualized teaching of mathematics

PHYLLIS ZWEIG CHINN

An associate professor of mathematics at Towson State College in Towson, Maryland, Phyllis Chinn teaches courses in mathematics content and methods of teaching. She has been exploring ways to teach mathematics effectively at all levels, from preschool through college. She has also served as mathematics consultant for teachers in several parts of the country.

In recent years many people have recognized the value of letting children learn by doing. Issues of the ARITHMETIC TEACHER have been devoted to mathematics laboratories and individualized teaching. Several books make eloquent pleas for using a child's natural curiosity and innate learning skills to allow him to continue to grow in all learning, including mathematics. *Crisis in the Classroom, Freedom to Learn,* and *What Do I Do Monday,* provide detailed statements of the weaknesses of some traditional approaches to teaching and the exciting possibilities inherent in some alternatives.

In a mathematics content and methods of teaching course that I teach at Towson (Maryland) State College, I have evolved some styles of teaching that may help potential teachers to develop self-confidence and skills for teaching mathematics (and all the elementary school curriculum) in more activity-oriented ways. The following detailed description of the format of the course in Methods of Teaching Mathematics in the Primary School may give ideas to teachers in the primary and elementary schools, as well as to those responsible for the training of teachers in the colleges and universities.

On the first day of class, students receive a brief description of activities for the whole semester. The several types of participation that are to be required of students for this course are outlined. Students are expected to—

- perform mathematical activities during class meetings,
- contribute to discussions,
- do independent reading,
- begin a file of ideas to be used in teaching mathematics,
- learn to use a "junk corner,"

and, finally, they must teach some children, using techniques learned in this course.

The remainder of the first class period is devoted to a fairly structured small-group activity. The students are divided into groups of three. Each group selects a project from a list provided. The projects are all different, but each project involves gathering data, organizing it, and displaying the results in at least six different ways, including a written summary and graphs. Ideas for these projects were selected primarily from those listed in suggestions for teacher workshops in *Freedom to Learn* (p. 89). The activities include comparing pulse rates before and after exercise, determining the pressure exerted by a person on the bottom of his feet, and comparing the height and reach of individuals. The students must decide how to gather data, how much data to gather, and how they will present the results. They begin the measurements during the first class meeting and continue them out of class.

The second class period is devoted to correlating and displaying the results of the measurements made by students. No letter grade is assigned to these projects, or to any other in-class activities of the students. The presentations are collected, however, and returned to students with extensive comments on the accuracy and suitability of the records made.

Following these first two days of directed activities, the class becomes an adult-level mathematics laboratory. The classroom is supplied with diverse learning aids, many of which can be found in the better equipped elementary school classrooms. For example, students have access to the following materials: Cuisenaire rods, plastic polyhedra, balances, geoboards, abaci, attribute blocks, dice, rubber bands, and duplicated pages of multiplication tables and hundreds charts. Ideas for using this equipment have been collected from many sources. Each idea is recorded on a separate "activity card." These activity cards tell what to do, how many people are to participate in the activity, and what equipment to use. The Midwest Publishing Co., Inc. has a series entitled "A Cloudburst of Math Lab Experimenting," which has some ideas for similar activity cards. Many of the activity cards are based on instructional manuals issued with the various pieces of equipment. Some are based on ideas from the three books mentioned in the first paragraph of this article.

A flow chart of the activities is posted; this shows which are relatively simple, which are more difficult, and which ones may give help in performing certain others. The flow chart is color coded to show which areas of mathematics are involved: geometry, measurement, basic operations, probability, or using elementary school equipment. Each student is expected to complete activities from each of these areas, and to try to become familiar with all of the equipment in the room. Students are free to make up their own activities at any time. By carrying out these activities, students potentially learn many

things: how it feels to learn through discovery, various ways a mathematics laboratory can be run, many ways to use the elementary school mathematical equipment, the role of the teacher in an open classroom, and the fact that people can learn on their own in a properly prepared environment. We hope that by working on laboratory exercises the students will develop self-confidence in handling mathematics problems; that they will learn to figure things out independently rather than depending on the teacher to say what to do, when to do it, and how to tell if it is done correctly.

After several periods of working in the mathematics laboratory, other experiences are provided for students. The Nuffield Project has produced several excellent films on learning mathematics, including "I Do and I Understand" and "Maths is a Monster." These films show mathematics classrooms in action. They are among several movies viewed by the students. During occasional discussions involving the entire class, students talk about the mathematics they see in elementary level classes they observe, things that take place in the methods-of-teaching class, things they have read, and problems they encounter. I participate in these discussions, but try not to lecture.

Considering how rapidly the elementary school curriculum is changing, it would be impossible to attempt to tell these students exactly what to teach or specifically how to teach it. Instead they are exposed to sources of ideas, and ways by which to evaluate teaching ideas. One source of teaching ideas is writings by teachers, and one of the out-of-class requirements for the course is reading. Each student is expected to do readings in areas concerning mathematics teaching and curricula, including the basic operations, numbers and numerals, geometry, measurement, individualized methods of teaching mathematics, Montessori, and Piaget. The readings may be from any sources, but they must include articles from the ARITHMETIC

TEACHER, other professional journals, and the teacher's guides to some new textbooks and workbooks. Certain books are recommended, including those mentioned in the first paragraph of this article and the Nuffield Project series.

Students are to take notes on their readings in whatever way will be most useful to them. They do not hand in their notes. Everything they are getting from the reading is reflected in their idea file.

Each student is required to compile a file of ideas for his own teaching. This file is to include ideas for teaching, reinforcing, and testing all of the areas of mathematics commonly taught in the elementary school (or those grades which the individual would ever consider teaching). The students must decide which areas to include on the basis of reading student texts, teacher guides, curriculum outlines provided by local school districts, or any other source. The file must be organized in a fashion that will make the ideas readily accessible for teaching. The ideas may come from anywhere—class activities, observing in elementary schools, readings, other students, and so on. The files are collected twice—midway through the semester and near the end. The files are graded; these are the only letter grades assigned to any student work. The files of ideas are evaluated on suitability of the ideas and materials involved, ease of extracting information, how interesting the ideas are, variety of types of activity, variety of mathematical ideas covered, and organization. At the end of the year, students are given copies of ideas for many activities, insuring that all students have some good ideas in their files.

An elementary school classroom should have a "junk corner," a place equipped with things a child may use in any way he pleases (except, of course, to hurt anyone). For our preservice teachers to see this idea in action, there is a junk corner in the college classroom. Things in this area may be taken by any student to be used in any way. Anyone who has access to extra things of any type is invited to contribute junk. Local stores contribute things they would ordinarily discard: packing materials, paper, film canisters, and boxes. Many things from around the home can be used for mathematics: egg cartons, empty boxes and cans, cylinders from paper towels, empty spools, macaroni, partial decks of cards, broken clocks—almost anything. In this era of increasing ecological concern, teachers accustomed to having a junk corner as a "recycling center" will see ways to use things that they might otherwise have discarded.

There is one other major project for this course. Near the end of the semester, each student prepares an activity for a mathematics learning center suitable for a young child. On a designated morning, the students set up a mathematics laboratory for children in a nearby elementary school. Thus the college students get to observe children learning mathematics by active participation. The children are free to choose any activity that is displayed; they tend to be delighted with the activities and to show a long attention span. From this experience, the potential teachers seem to realize that young children can learn in a manner similar to that in which they have been engaged during the semester. We hope that many of the prospective teachers leave this class prepared to try laboratory teaching in their own classes.

References

Biggs, Edith E., and James R. MacLean. *Freedom to Learn*. Don Mills, Ont.: Addison-Wesley, Ltd., 1969.

Buckeye, Donald A., et al. *A Cloudburst of Math Lab Experiments*. Troy, Mich.: Midwest Publishing Co., Inc.

Holt, John. *What Do I Do Monday*. New York: E. P. Dutton & Co., Inc., 1970.

Silberman, Charles. *Crisis in the Classroom: The remaking of American Education*. New York: Random House, 1970.

Topics from Elementary School Mathematics

TEACHING how to teach concepts, rational numbers, geometry, and problem solving is the focus of this section. A recurring theme is that these elementary school topics should be taught to preservice teachers in the same way as they should be taught to children.

In "Helping Preservice Teachers to Teach Mathematical Concepts," Davis outlines a lesson format based on six types of questions, which he gave to his students to use when peer teaching and microteaching.

Lester argues that teachers' inadequate knowledge about rational numbers is a major reason for the difficulties children have, and he proposes in "Preparing Teachers to Teach Rational Numbers" that both children and their prospective teachers should be taught the concepts in a similar fashion.

Two articles focus on geometry, another area in which teachers often have little background. In "Topics in Geometry for Teachers—a New Experience in Mathematics Education," Kipps describes a course based on an active learning approach, in which university classes were conducted in the same way that corresponding classes ought to be taught in elementary school. Wong also used geometry to provide prospective teachers with the opportunity to learn through an informal, inductive, manipulative approach at levels appropriate for teachers. She presents four examples illustrating the approach in "Geometry through Inductive Exercises for Elementary Teachers."

Problem solving is dealt with in four articles, all based on the premise that before children can become good problem solvers, their teachers must be good problem solvers themselves. In "Simulating Problem Solving and Classroom Settings," Liedtke and Vance describe the simulation of problems and instructional settings that preservice teachers might in turn use with children. Zur and Silverman identify a level of mathematical thinking they call the "open search" approach and illustrate it with a problem and detailed suggestions for using this approach with preservice teachers in "Problem Solving for Teachers." Krulik and Rudnick outline a flow chart of the problem-solving process in "Teaching Problem Solving to Preservice Teachers" and explain how to use the chart in providing the same kinds of activities and experiences for preservice teachers that they might provide their own students. In "Elementary Teacher Education Focus: Problem Solving," LeBlanc draws attention to the importance of having preservice teachers experience success in problem solving while they are learning how to teach the process to children.

Teacher Education

Helping Preservice Teachers to Teach Mathematical Concepts

By **Edward J. Davis**

University of Georgia, Athens, GA 30602

Teaching mathematical concepts is important. In a time when students' performance on basic skills is being emphasized, teachers must be alert not to stress computation at the expense of concepts. Understanding the meaning of a fraction differs from knowing how to add fractions. A lack of conceptual knowledge of a fraction can contribute to incorrect performance in computational skills with fractions. The same can be said about other mathematical concepts, such as percent, decimals, division, factoring, like terms, square root, and so on.

What Are Mathematical Concepts?

Simply put, a mathematical concept is a collection of meanings that one associates with a word used in mathematics. The set of associations you have with the term *altitude* constitutes your concept of altitude. These associations or meaning did not develop all at once, and they will become more numerous as you learn more about altitudes.

Open any school mathematics textbook to the index and begin reading. You will be bombarded with mathematical concepts and the associations concerning these concepts that are introduced, reviewed, and extended throughout the text. The main headings of the index of the text that I have before me begin like this:

Addends
Addition
Algorithm

Area
Attribute Pieces
Center
Centimeter

Six of these first seven entries are mathematical concepts.

We are teaching a mathematical concept whenever we teach the meaning of a word used in mathematics. The extent of a student's conceptual knowledge is probably a strong indicator of this student's performance in class, on tests, and in applying mathematics to solve problems.

How Can We Teach Mathematical Concepts to Children?

I see at least six main tools, or moves, that mathematics teachers can use to help students create a well-balanced set of associations with any given word in mathematics. They are presented in the following example that deals with the concept of *percent*. The teaching tools, or moves, are generated from six types of questions that can give evidence of a student's knowledge of percent.

Question type (setting)	Teaching tool (move)
1. Can you give some examples of 30 percent, 60 percent, 100 percent, and 150 percent of these objects or collections?	Examples
2. Which of these pictures definitely does *not* show 25 percent of the object or collection? Why not?	Nonexamples
3. Think about a can of Coke. About what percentage of the can would someone drink if they took— a sip? a gulp? half the can? almost the whole can? the whole can? two cans and a sip more?	What's True
4. Rex claims that he has cut off 30 percent of this string. How can we check to see?	Testing (Guarantee)
5. What is the difference between 40 percent and 80 percent of this segment? How is 120 percent of a shoelace like 20 percent of the shoelace?	Compare-Contrast
6. What do you think the word *percent* means?	Definition

Keep in mind that each of the six settings listed on the left is just one possible way of helping students acquire the corresponding kind of association or meaning labeled on the right. Students can be said to have an understanding of a concept to the extent that they can successfully make or respond to moves involving Examples, Nonexamples, What's True, Testing, Compare-Contrast, and Definition that are made or posed by the teacher. These moves can be spontaneous questions or statements or those made in response to students' questions. Posters or bulletin boards frequently display situations involving Examples or What's True. Tutorial sessions at a computer frequently draw from these six kinds of understanding about mathematical concepts.

Training Preservice Teachers to Teach Concepts

To the extent that course evaluations by student teachers are valid and that my observations of lessons taught by them are accurate, I have had success in training preservice teachers to teach mathematical concepts by using peer teaching and microteaching. The lessons are ten to twenty minutes in length. Preservice teachers are urged to proceed according to the following format whenever possible.

1. Set an objective for the lesson, motivate students, and review the appropriate knowledge.

2. Present four or five *examples* of the concept—make them significant examples. Try to use some concrete representations.

3. Present two or three *nonexamples* of the concept. Use some nonexamples that are reasonably close to being examples and point out why they are not examples.

4. Attempt to get the students to identify *what's true* about the concept through the use of your examples and nonexamples (it is hoped that they are still in view).

5. Have students *practice* identifying and classifying situations as examples or nonexamples of the concept.

6. Discuss students' selections and identify those *tests* that help students decide if an object is an example.

7. If feasible, *compare* this concept to a related concept known to the students or *contrast* different examples of the concept to show similarities and differences.

8. Have students *define* the concept using their own terminology. Consider comparisons to the textbook definition.

9. Provide more practice for the students.

A few years ago, the thought of prescribing for teachers a sequence of steps for a lesson plan was disturbing to me. I felt that preservice teachers should use their own creativity and artistry in sequencing the lesson on concept development. Nonetheless, they seem more successful with this new format than with the nondirected approach. Let me add a few more observations:

- Students are to follow it three or four times and then begin to change it according to their beliefs and experiences.
- Research by Dossey (1976), Shumway (1974), and others indicates no advantage in any particular sequence for teaching mathematical concepts.
- If the lesson does not proceed as smoothly as students would like, they can share the "blame" with the instructor.

Prescribing a lesson format may keep preservice teachers from being overwhelmed by theory and give them some security. Student teachers deliberately make changes in this format within the span of a few lessons and are sometimes "forced" to depart from it in response to students' questions. These variations are good. The prescribed format, although appearing to control teaching behavior, still leaves the teacher considerable latitude in the choices for examples, nonexamples, and physical referents used in the presentation and in the practice.

References

Davis, Edward J. "A Model for Understanding Understanding in Mathematics." *Arithmetic Teacher* 26 (September 1978):13–17.

Dossey, John A. "The Role of Relative Efficacy Studies in the Development of Mathematical Concept Teaching Strategies: Some Findings and Some Directions." *Teaching Strategies: Papers from a Research Workshop.* Columbus, Ohio: ERIC/SMEAC and Ohio State University, 1976. (ERIC Document Reproduction Service No. ED 123 132)

Shumway, Richard J. "Negative Instances in Mathematical Concept Acquisition: Transfer Effects between the Concepts of Commutativity and Associativity." *Journal for Research in Mathematics Education* 5 (November 1974):197–211.

Teacher Education

Preparing Teachers to Teach Rational Numbers

By **Frank K. Lester, Jr.**
Indiana University, Bloomington, IN 47401

Other articles in this issue present valuable ideas and points of view about learning and teaching concepts and skills involving rational numbers. In particular, these articles point out the trouble spots for children, offer suggestions for teaching these difficult topics, and offer some useful suggestions about sequencing instructional activities. An underlying theme of these articles is that rational-number topics are difficult to teach and difficult to learn. I propose that *a major reason why elementary school children find rational numbers so troublesome is that some of their teachers have an inadequate understanding of rational-number concepts and just as poor facility with rational-number skills.* Because of this lack of understanding and skill, teachers often cannot decide how to present these topics to children and which ideas to emphasize. Too often, teachers resort to little more than rote-level instruction. In this article, I offer some suggestions for resolving this problem during the mathematics education training of prospective teachers.

Prospective Teachers' Knowledge of Rational Numbers

During my career as a teacher educator, I have taught mathematics methods courses to well over 1000 preservice elementary teachers. Anyone who has taught such a course is well aware that as a group, elementary teachers' understanding of mathematics is weak. Furthermore, their mathematics deficiencies are compounded by generally negative attitudes about, or even anxiety toward, mathematics. The inadequacy of elementary teachers' mathematical knowledge is nowhere more apparent than in their understanding of rational numbers. Let me document this statement.

For the past five years all preservice elementary teachers at Indiana University have been required to pass an arithmetic competency test to receive credit for the first of three three-credit-hour courses in mathematics for elementary teachers. Items on the test assess mastery of arithmetic concepts and skills at no higher

---

Rote learning leads to rote teaching.

---

than a seventh-grade level, and most items require no more than fifth- or sixth-grade knowledge of arithmetic. To pass this test a student must correctly answer approximately 75 percent of the items in each of seven categories (numeration, arithmetic operations, fractions and decimals, percentages, signed numbers, ratio and proportion, and word problems). Of the more than 600 students who have taken the test to date, approximately 50 percent have failed on the first attempt to reach the 75 percent criterion for passing. Failure most often results from poor performance in the rational-number categories. To learn more about the nature of their difficulties, I have talked privately with students who have failed the test. A typical conversation usually proceeds as follows:

Teacher: You failed because you missed too many of the items on fractions and decimals. Did you just have a bad day, or don't you understand how to do these problems?

Student: Well, I think I understand them pretty well, but I just don't do very well on math tests. Yeah, I had a bad day.

T: That's possible, we all have off days from time to time. I want to help you prepare for the next test so you will do much better on the sections you failed. To help you I need to know more about why you got the problems wrong. For example, you wrote

$$\frac{56}{21} \div \frac{2}{7} = \frac{3}{28}.$$

How did you do this one?

S: Let's see. Well, I remembered that you have to "invert and multiply" when you divide fractions, so I inverted $\frac{56}{21}$ and got

$$\frac{21}{56} \times \frac{2}{7}.$$

Then I did some canceling and finally multiplied to get $\frac{3}{28}$.

T: Do you see why that's wrong?

S: I guess I inverted the wrong part. It must be

$$\frac{56}{21} \times \frac{7}{2}.$$

Right?

T: That would be one correct way to do it. But what I'd like to know is why did you think to "invert and multiply"?

S: That's what I was told in fifth grade or whenever it was. I guess I just forgot which one to invert.

T: So you think that if you brush up on rules such as "invert and multiply" you'll pass next time?

S: I think so, because I really didn't take this test seriously. I used to know this stuff, I just forgot.

T: But you didn't forget how to do

$$347 \atop \times 86$$

Why is that?

S: It's easier. Anyway, it seems like there are so many rules to remember when you work with fractions. It's hard to keep them straight.

T: OK! Let me ask you another question. Do you think you could teach fractions to fourth or fifth graders?

S: Well, I do need to brush up, but I think I could.

T: Would you enjoy it?

S: I really like working with kids. I wouldn't really *enjoy* teaching fractions, but since I like teaching, it would be OK. My biggest problem would be not to make it boring. Fractions were boring for me in school. In fact, that's one reason I started hating math.

T: Because math is boring or because fractions are boring?

S: Both! By the time you're in fifth or sixth grade it seems like all you do in math is fractions and decimals and things like that. They are boring, so math was boring.

T: Was it boring because it was hard or because it wasn't interesting?

S: I'm not sure, but probably both.

T: I wouldn't think I would enjoy teaching a topic I thought was boring. I'll bet the kids would think it was boring, too.

S: Probably so! To be honest, I don't plan to teach the upper grades. I want to teach grade one or two.

T: What if you don't have any choice and you are offered a job in a good location, but it's for sixth grade?

. . .

The point of this simulated conversation is to show that students' failures on rational-number exercises often stem from their having learned by rote a collection of disconnected rules and procedures. Furthermore, because these rules and procedures were not practiced regularly, they were easily forgotten—a striking illustration of the limitations of rote learning. In addition, instruction that leads to mechanical learning will probably be repeated when these students become teachers. As undesirable as rote learning is, it is especially so for a prospective teacher. *An individual who has only rote-level mastery of a topic cannot be expected to guide others to any more than rote-level mastery of that topic.*

Overcoming the Difficulties

Teachers of elementary mathematics methods courses are faced with a difficult question: How does one instruct students how to teach topics that they neither like nor understand? The answer would seem to have at least three parts. First, students' competence in working with rational numbers must be improved. Second,

---

Teaching preservice teachers about rational numbers should follow the same approach as their teaching with their students.

---

teacher educators must provide their students with conceptually sound ideas about the sequencing of topics, use of instructional aids, introduction and development of new topics, applications of concepts and skills, and other aspects of instruction in rational numbers. Finally, students' attitudes must be changed by reducing their anxieties about rational numbers and convincing them of the value of meaningful instruction.

Unfortunately, most programs for teacher education are designed so that these changes must occur in a few class meetings. To expect any significant improvements, prospective teachers should be taught the content of rational numbers in a manner similar in *development* and *emphasis* to the way they should teach elementary schoolchildren. The advantage of this approach is that the issues of content competence and knowledge of pedagogy can be addressed simultaneously. Furthermore, the study of mathematics content in relation to methods of teaching it to children should increase its relevance to the prospective teacher. As the content becomes more relevant, motivations to learn and attitudes improve. Let us now consider the appropriate development and emphases of this approach.

Development of Rational-Number Concepts

Preservice teachers must recognize the importance of the development of rational-number ideas. A major reason for children's difficulties with rational numbers is that insufficient attention is paid in the primary grades to the establishment of a foundation on which to build further understanding and skills. Instruction in rational numbers for preservice teachers should begin with fundamental conceptualizations and proceed systematically to computational skills in, and their application to, various problems. Both preservice teachers and children should have enough exposure to models of rational numbers to allow them to form meaningful mental representations. If this approach is taken, the symbolic representation of rational numbers becomes linked to real situations and concrete models.

Emphases in Teaching Rational Numbers

To be effective, teachers must know what topics should be emphasized during instruction. Four basic categories of emphasis include (1) the *development of rational-number ideas* from real-world situations to symbolic forms; (2) the *modeling* of rational numbers; (3) the use of *big ideas* in developing comprehension of, and skill with, rational numbers; and (4) the importance of *systematic analyses of children's error patterns*. To illustrate each category, let us discuss them with respect to fractions.

Development of rational-number ideas. The study of fractions should begin with examples from the children's lives. Models of fractional ideas, both concrete and pictorial, should be developed from these examples to aid the children in the formation of meaningful mental representations of fractional concepts. Real-world examples and models give meaning to both the symbolic form and the oral name for fractional numbers. If too little attention is paid to the establishment of solid connections among models, symbols, and names, children will have difficulty when more abstract work begins.

Models of rational numbers. Several models should be used to introduce the concept of fractions. The most common model, and one that should be introduced first, is the *region model*. Various examples of regions should be used—rectangular, circular, and so on. Region models are effective in giving children a grasp of fundamental concepts about fractions. A second model is the *set model*. Whereas the region model depicts a whole unit divided into parts, the set model illustrates a part of a set of discrete objects. (This model is often more difficult to understand.) A third model is a *measurement model*, an example of which is a number line. Number lines are often effective in introducing equivalent fractions. Models are very important in establishing a base on which to build later concepts; it is important that fractions be modeled in different ways—with regions, sets, and number lines. The region and set models are discussed elsewhere in this issue. Some references on measurement models are included at the end of this article.

Big ideas. At least four big ideas must be stressed to develop children's competence with fractions. The first has already been discussed—the importance of establishing stable connections among models, symbols, and oral names. The notion of equivalent fractions is another big idea because it is critical to the further development of ideas about rational numbers. The third big idea involves the development of techniques for comparing and ordering, performing operations with fractions, and changing fractions to equivalent forms. This idea receives the most attention, often at the expense of the other big ideas. The fourth big idea involves the relationships among fractions, decimals, and percentages. Unfortunately, these relationships often are not emphasized.

Analysis of error patterns. Providing prospective teachers with experience in analyzing samples of children's work to identify common errors is effective in sensitizing teachers to common difficulties that may arise. Once these trouble spots are identified, teachers can see why certain ideas and skills must be presented carefully.

Summary

The premise of this article is that teachers' lack of knowledge about rational numbers is a major reason for the difficulties children experience in learning rational-number concepts and skills. Often, teachers' inadequate knowledge results in the presentation of rational-number topics as a series of disconnected ideas and procedures with little or no meaning. My solution to this problem is to teach rational numbers to prospective teachers in the same way as the concepts should be taught to children. Not only must teachers be taught how to teach rational numbers but also they must understand what they are teaching; they need to be confident that the instruction they are providing is promoting solid understanding of concepts, reasonable facility with skills, and attitudes conducive to subsequent learning.

Bibliography

Ashlock, Robert B. *Error Patterns in Computation*, 2d ed. Columbus, Ohio: Charles E. Merrill Publishing Co., 1976.

Kennedy, Leonard M. *Guiding Children to Mathematical Discovery*, 3d ed. Belmont, Calif.: Wadsworth Publishing Co., 1980.

LeBlanc, John F., Donald R. Kerr, G. R. Croke, K. M. Hart, C. J. Irons, and Thomas L. Schroeder. *Rational Numbers with Integers and Reals*. Reading, Mass.: Addison-Wesley Publishing Co., 1976. ◗

Forum on teacher preparation | *Francis J. Mueller*

Topics in geometry for teachers—a new experience in mathematics education

C A R O L H . K I P P S
University of California, Los Angeles, California

Carol Kipps, besides teaching courses at
the University of California, Los Angeles, is highly active with in-service
projects, among them the Madison Project, the California Conference
for Teachers of Mathematics, and UCLA Extension.

Can teachers capture by themselves the excited enthusiasm shown by children in classes sponsored by such curriculum groups as the Madison Project or Nuffield Project? Can a teacher reared on lecture-drill-homework classes feel and show the drama inherent in "I do and I understand" activities, in peer-group discussions, and in concepts such as the concrete-ikonic foundation of abstraction? A new course at UCLA is focusing on these dynamic factors so that teachers will know their value from their own personal experiences and feelings.

Geometry is the vehicle, and grades K–8 is the level. Geometry was chosen because prospective teachers have little background in geometry and very often fear having to teach it. More and more geometry is being introduced in the elementary grades. Teacher-training research and the recommendations of professional organizations show that geometry is more troublesome than arithmetic or algebra.[1]

Modern curricula aimed at optimal sequencing capitalize upon the child's early curiosity about shapes, the relations between shapes, and patterns. Informal exploratory geometry provides the necessary basis for later symbolization and abstraction. Also, an active learning approach requires a different kind of teacher behavior. When small groups of students are involved, the role of the teacher is more

---

1. See for example: *Goals for Mathematical Education of Elementary School Teachers: A Report of the Cambridge Conference on Teacher Training* (Boston: Houghton Mifflin Co., 1967). *Course Guides for the Training of Teachers of Elementary School Mathematics,* rev. ed. (Berkeley: Committee on the Undergraduate Program in Mathematical Association of America, 1968). Carol Kipps, "Elementary Teachers' Ability to Understand Concepts Used in New Mathematics Curricula," THE ARITHMETIC TEACHER 15 (April 1968): 367–71. Marilyn Suydam, "Research on Mathematics Education, Grades K–8, for 1968," THE ARITHMETIC TEACHER 15 (October 1968):531–44.

demanding—and far more rewarding. As the teacher moves from group to group listening to the dialogue, she must consider when to ask a question, when to be silent, and when to withdraw altogether.

Many people tend to teach the way they have been taught. This can be a virtue as well as a hazard. In the experimental class taught at UCLA during the winter quarter of 1969 and then repeated in the summer, the class was conducted in the same way that corresponding classes ought to be taught in the elementary school. For not only can the process be modeled, but the teacher can evaluate it from personal experience, choosing appropriate learning activities and peer groupings with greater insight and precision.

Method

The course began with a discussion of the goals pupils should achieve by the end of the eighth grade. These behavioral objectives free the teachers from complete reliance on the basic text and focus on individualizing the learning opportunities to achieve at the 100-percent level. The following objectives were suggested as basic and minimal.

SHAPES IN GEOMETRY

1. The child will name flat or space figures when shown a physical model or a pictorial representation of the following: triangle, quadrilateral (square, rhombus, trapezoid, parallelogram, rectangle), circle, ellipse, cube, cone, rectangular solid, sphere, prism, pyramid.
2. The child can show where he would measure a flat or space figure to find the length of its diagonals and its altitude. Also, the child can draw a line on a pictorial representation to indicate what he would consider the altitude or a diagonal of the figure to be.
3. The child will state whether flat or space figures have point (turning) symmetry or line (folding) symmetry and define these ideas.
4. Given a physical model, a picture, a verbal description, or a description in set language, the child will state whether the figure is open or closed and whether it is convex.
5. The child will construct a physical model or sketch and will describe essential properties of triangles and tetrahedrons; squares, rectangles, cubes, and rectangular prisms; circles, cylinders, cones, and ellipses; polygons, regular plane, and space figures.
6. The child can name, define, and represent the foundation elements of geometry: points, lines, line segments, rays, planes, and angles.

RELATIONS BETWEEN SHAPES

1. The child will pick out congruent shapes and verify his decision by fitting. He will define congruent figures as those that can be made to fit together and use the notation $\cong$ for congruent.
2. The child will classify from a set those figures that are similar.
3. The child can identify examples, list examples, and sketch examples of the following relations between geometric shapes: covering (tessellate), separating, inside, outside, on, and topologically equivalent.
4. The child will identify, construct, sketch, or use set notation to describe the possible intersection sets for lines; lines and plane regions: lines and solid figures; two plane regions; and a plane solid figure.
5. The child will locate a given point on the number line. Given a pair of coordinates (x, y) that belong to the set of rational numbers, the child will locate the given point on a number plane.

MEASUREMENT

1. Given practical problems involving measurement, the child will experiment (estimating, selecting appropriate units, gathering data from observations, constructing number scales, and computing) and will attempt to verify his solution in some fashion.
2. Given a standard unit, the child can roughly approximate and measure the length, area, and capacity of common objects in appropriate English and metric units. The allowable margin of error will depend on instruments used, if any. Since all measurements are approximate, the child can cite methods for reducing errors of measurement.
3. The child will verbalize that measurement is the process of selecting an appropriate standard unit and finding a number that compares the two. The child will state generalizations that follow, based on measurements and the tables or graphs in which the measurements are recorded.

LOGIC AND PATTERNS

1. Given one, two, or three criteria for selection, for example, color (e.g., red), shape (e.g., square), and size (e.g., large), the child can classify elements of a set in a way that shows which elements have none, one, two, or three attributes.
2. The child can determine the pattern of a sequence of figures or numbers, continue the sequence for at least three more elements, and verbalize the criteria that he is using.

The big change in method in these experimental classes was that the university students were asked to work together in small groups on learning opportunities designed for specific behavioral objectives. What lecturing there was had to do mostly

with learning theory—Piaget, Bruner, Gagné. While solving the problems, the students were encouraged to use many kinds of resources, e.g., concrete embodiments of various mathematics ideas such as Dienes Multibase Arithmetic Blocks, Cuisenaire rods, and geoboards. A curriculum library was available that included dictionaries, textbooks, and teacher guides from the experimental projects such as Nuffield and Madison Projects as well as those for standard texts. The instructor acted as a consultant, answering questions with questions and suggesting references. As discussions waxed about the definitions of words such as *diagonal* and *altitude,* their adequacy for the present two-dimensional problems or similar ones in three dimensions were debated. Preservice teachers enjoyed "playing teacher" with each other and using clues to draw out those more naive mathematically. In an in-service situation, the instructor would have many "assistants."

A second key difference from the usual university course in education was the use of evaluation as one of the learning activities. As part of the course, the students not only answered questions, but proposed them! Many educational taxonomies suggest that asking an insightful question aimed at a particular learning behavior is a much higher cognitive skill than answering questions. After solving a mathematics problem in their peer-group situation, students were asked to devise a suitable test item. At this point each person was asked to make an individual contribution, but by all means to consult with the group about it. While the test item was to be based on the stated objective, it might well include prior skills or knowledge and need not be a paper-and-pencil type of test item.

To illustrate this classroom activity, the following is a page taken from the class notes. The students were required to make their own geoboards and to bring them to class for this type of activity. First, as a teacher, each was to read the specific behavior objective. Next, as a pupil, each had to solve the problem, discussing it with members of his small group. Last, again in the role of a teacher, each participant was to read the sample test item and write another of his own based on the objective.

SAMPLE CLASS WORKSHEET

Specific behavioral objective.—Identify the diagonals of various plane figures and define the idea of a diagonal.

Learning opportunity.—Make the following shapes on your geoboard. Transfer each figure to dot paper and draw in all the diagonal lines using a red pencil. In which shapes are the diagonals of equal length?

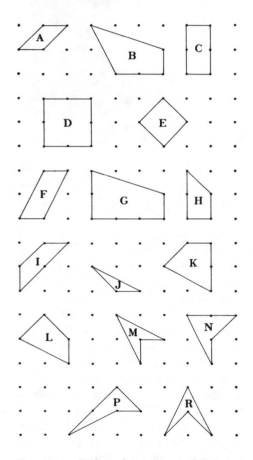

Test item.—Define what you mean by the term diagonal. Draw a shape that has no diagonal and tell why it doesn't by applying your definition.

Test item.—Does a diagonal necessarily bisect the angle at that vertex?

Comments.—A follow-up discussion might develop the idea of whether the definition given

would work for space figures, or for a line joining two vertices which is not a side.

Evaluation

This methods course in mathematics education at UCLA is organized primarily to develop—

(1) skill in planning and evaluating learning opportunities in mathematics for pupils;

(2) skill in using the Socratic approach;

(3) knowledge and skill in applying basic concepts of informal geometry.

At frequent intervals, test items written by the students on the worksheets were reviewed by the instructor and returned with suggestions or comments. Grades were not given for these worksheets, because it was hoped that the content and experiences would foster interest in the learning activity rather than in some extrinsic reward. Relevance to the specific objective and mathematical correctness were checked. Creative style and elegance were noted with positive comments. It was a matter of considerable delight and astonishment to the instructor when not one member of the class of 44 asked about getting a "grade." This attitude is most appropriate to a study of teaching mathematics, for the "new math" was introduced not merely on the claim that it represented more important content but equally on the argument that it would build a new spirit of inquiry and creativity.

Grades for the course were assigned on the basis of an examination focusing on the methodology. Here is a sample question:

> Select one of the objectives from above and describe three learning experiences that you could provide to enable children to achieve the objective. Write one learning experience for each of the following levels:
>
> Enactive (sensory-motor)
>
> Ikonic (representational)
>
> Symbolic (abstract)

Continual feedback of geometric content was possible during the class periods, since the instructor could spend time with a small group or an individual student at no expense to the rest of the class. It is the rare student who will display his ignorance in a conventional classroom; but in the small-group approach, important questions are readily raised and discussed. When a student works in a small group, he finds it much easier to express his confusions enough to enable others to help him.

A term paper on a mathematics topic selected by the student was required by the course. The basic text was broadly representative of arithmetic, algebra, and geometry. Prior to the use of the geometry notes and discussion "in fours," few students had selected a topic from geometry for this paper. It is of note, then, that 28 out of the 44 felt comfortable with geometry and did their paper in this area.

Conclusion

Prospective teachers have little background in geometry and very often fear having to teach it. It is apparent that many potential teachers exposed to this approach will come away with a good feeling about geometry, some of the confidence needed to teach geometry, some exploratory activities they can use with children, and a much broader knowledge than that normally obtained from a straight geometry or mathematics course.

Some very important questions cannot yet be answered. Does this approach to teacher training tend to make the content easier to retrieve and to reconstruct? Will teachers include geometry in their lessons and use appropriate manipulative apparatus and resources beyond the pupil text? A follow-up is planned in the form of a questionnaire at the end of the first teaching assignment.

Student evaluations of the format were enthusiastic, and they expressed attitudes toward studying mathematics as well. One student wrote, "I think you should keep the course as it is, because the workshop atmosphere is what is needed in the classroom, and we as prospective teachers should practice in the same atmosphere."

Geometry through inductive exercises for elementary teachers

RUTH E. M. WONG

Currently an associate professor in the Department of Mathematics at the University of Hawaii, Ruth Wong has worked with preservice and in-service teachers.

Geometry frequently suggests only a sequence of definitions, postulates, theorems, and proofs. While this is part of geometry, it represents the finished product —a systematic, well-organized body of knowledge about plane and space figures. Preceding this formal organization into a deductive structure are a number of processes. These include: searching for patterns, recognizing patterns, making guesses, observing similarities and differences, assembling data, generalizing from specific instances, checking conjectures, expressing ideas accurately, and recognizing concepts in concrete situations (Hawaii Curriculum Conference 1966).

It is my view that experiences that allow pupils to engage in these processes should make up the geometry program in the elementary school. Rather than drill in geometric vocabulary or identification of figures from definitions or an early introduction of axiomatic geometry, there should be provision for visual and manipulative experiences that lead to geometric facts and relationships.

Suggested approach to teacher preparation

This view has definite implications for teacher preparation in geometry, the major one being that the teacher should have *personal involvement* with the very mathematical processes through which he will be providing experiences for his students. It is essential that, for the teacher as well as the student, geometric content be acquired through an informal, inductive, manipulative approach. This is not to suggest that the teacher simply study materials prepared for elementary pupils or that he do the activities that are expected of elementary pupils. It is to say that teachers should be provided opportunities to engage in the processes with geometric content listed above and at levels *appropriate for teachers*. The main emphasis should be to involve teachers in situations in which they are allowed to arrive at, verify, and verbalize their own conclusions rather than to present them with the conclusions already formulated by the textbook or the teacher.

The following examples illustrate this approach.

Example 1. How many lines are determined by a given number of points, no three of which are on the same line?

Given two points, we know that there is exactly one line through them. Add a third point (fig. 1).

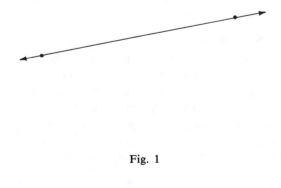

Fig. 1

How many more lines are there? How many lines altogether? Add a fourth point, being sure that no three are in a line (fig. 2).

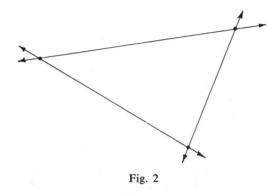

Fig. 2

How many more lines are there? How many lines altogether? Complete the table as you proceed (table 1).

Table 1

No. of points	No. of additional lines	Total no. of lines
2	—	1
3		
4		
5		
6		
10		
20		

Add a fifth point, with no three in a line. How many more lines are there? How many lines altogether?

If you were to add a sixth point, how many more lines would there be? How many lines altogether?

Now suppose you were given ten points.

a) How many more lines are there than for nine points?

If you knew the total number of lines for nine points, then it could be added to the number in *a* above to give the total.

b) But how many more lines are there with nine points than with eight points?

c) And how many more lines with eight points than with seven points?

d) How many more lines with seven points than with six points?

e) From the table, the number of lines for six points is ——.

So, the number of lines determined by ten points is the sum of *a,b, c, d,* and *e,* or —— + —— + —— + —— + ——, or a total of —— lines.

Now find the number of lines determined by twenty points.

Example 2. Draw a triangle.

Recall that the line segments determined by joining each vertex and the midpoint of the side opposite (medians) all meet at one point. Label this point *O* (fig. 3).

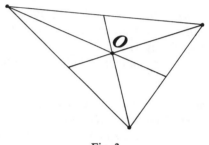

Fig. 3

Recall also that the perpendicular bisectors of the three sides all meet at one point. Label this point *P* (fig. 4).

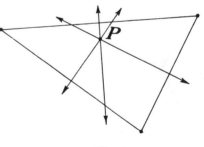

Fig. 4

Finally, recall that the lines through each vertex and perpendicular to the opposite side (lines containing altitudes) all meet at one point. Label this point *Q* (fig. 5).

Suppose we look at all three points together as in figure 6.

What do you notice about *O, P,* and *Q?*

Repeat the steps above for two more triangles, including an obtuse triangle. Do you find the same relationship among the three points?

What do you notice about $\overline{OP}$ and $\overline{OQ}$?

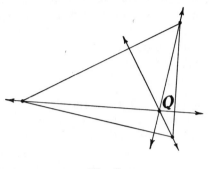

Fig. 5

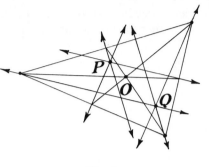

Fig. 6

What statement do you think can be made about the points of intersection of the medians, perpendicular bisectors, and altitudes of any triangle?

See if your statement holds for some other triangles.

Example 3. You have been provided with a set of construction-paper cutouts of polygonal regions. Examine them carefully, noting how they are alike or different. Then see how many different ways you can classify them. If possible, find at least one way that will divide the regions into two groups and at least one way that will divide them into more than two groups.

Note that among the possibilities are classifying the regions by number of sides, shapes, congruence, areas, convexity, and similarity. Names and definitions are provided only after distinctions among various types of regions are clear.

A few of the possible polygonal regions are shown in figure 7. Among the pictured regions, for example, regions 1, 2, 4, 6, 7, 8, and 9 can be observed to share a common property not found in regions 3, 5, and 10 (without prior knowledge of convexity).

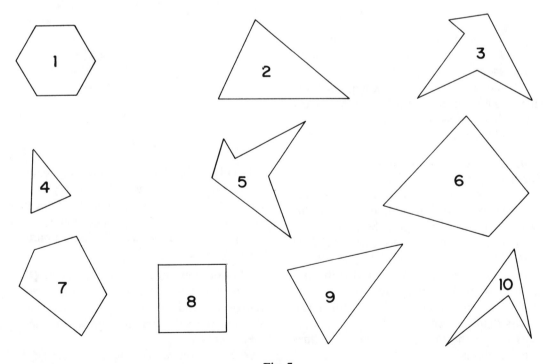

Fig. 7

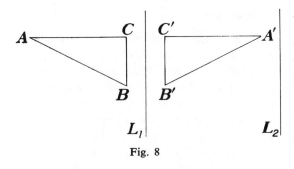

Fig. 8

Example 4. This exercise follows work with reflection congruent figures in which tracing paper is used to find images of figures under a reflection in a given line, and properties of reflections are studied.

What is the result of following one reflection by another reflection? Since two lines in a plane may be parallel or inter-

B' with B, and C' with C. (See fig. 8.)

Now reflect $\triangle A'B'C'$ in line L_2. Label this image $A''B''C''$, letting A'' correspond with A', B'' with B', and C'' with C'. (See fig. 9.)

How are $\triangle A''B''C''$ and $\triangle ABC$ related? Are they congruent? Are they reflection congruent? Why or why not? Using tracing paper, trace $\triangle ABC$. What can you do with the tracing paper to make the tracing of $\triangle ABC$ fit on top of $\triangle A''B''C''$?

Now draw another figure F on the sheet with the parallel lines, L_1 and L_2. Reflect this figure in line L_1, and label the image F'. Then reflect the image in L_2. Call the last image F''. (See fig. 10.)

How are F and F'' related? Are they

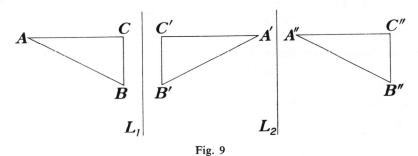

Fig. 9

secting, the axes of reflection may be parallel or intersecting lines. Let us first use parallel axes, L_1 and L_2. Using tracing paper, find the image of $\triangle ABC$ under a reflection in line L_1. Label the image $A'B'C'$, making sure that A' corresponds with A,

congruent? Are they reflection congruent? Are they related in the same way as $\triangle ABC$ and $\triangle A''B''C''$?

Using tracing paper, trace figure F and $\triangle ABC$. Repeat what you did above with the tracing paper to make the tracing of

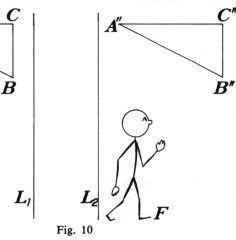

Fig. 10

$\triangle ABC$ fit on top of $\triangle A''B''C''$. What has happened to the tracing of F?

Try the sequence of two reflections with a third figure.

Make a statement about the result of performing one reflection in a line followed by another reflection in a parallel line.

Experimental classes

This approach was tried out with small classes of prospective elementary teachers at the University of Hawaii for four semesters in 1969–71. Six units of geometric content were selected—congruence, geometric inequalities, parallelism in planes and space, polygons and polyhedrons, area and volume, and transformations in the plane. For each unit, informal exploratory activities served as a basis for drawing conclusions. This was followed with a discussion of conclusions drawn and exposition of major ideas involved, including proofs when appropriate.

A limited testing program was designed for two groups composed of eighteen and twenty-four students. There were significant gains in knowledge of geometric content reflected by scores on a multiple-choice geometry test, more positive feelings toward geometry on an attitude scale, and greater ability to utilize geometric knowledge in tasks related to teaching elementary pupils (Wong 1971).

Of greater significance than test scores were the comments from the classes. Free response to a request for comments on the course was typified by the following: "This exploration and experimentation has been a pleasure—the result will be far more long-lasting." "I now feel more comfortable and confident in the area of geometry." "The discussions and exercises we do make us think about what we are doing." "This is the first class where I've seen 'discovery learning' in action and actually participated." "The class was good because of direct involvement of the student."

Summary

If elementary teachers are to be encouraged to provide their pupils with opportunities to deal with geometric concepts in an informal, inductive, manipulative manner, it is essential that the teachers themselves be given these kinds of experiences at their own level in their preparation program. One such approach to geometry for teachers has been suggested.

References

Hawaii Curriculum Conference. "Report of the Mathematics Group." Mimeographed. Honolulu: Hawaii Curriculum Center, 1966.

Wong, Ruth E. M. "Report on Experimental Geometry Course for Elementary Teachers." Mimeographed. Honolulu: University of Hawaii, 1971.

Simulating Problem Solving and Classroom Settings

By **Werner Liedtke** *and* **James Vance**

The simulation of various classroom settings in which problem-solving activities occur is one aspect of the mathematics methods course for elementary teachers at the University of Victoria. By participating in these sessions, the students are made aware of some of the possible instructional settings that they might use as teachers and the important role that problem solving plays in the curriculum. Some of these settings and problems are described here and some results of the students' involvement in the program are illustrated.

The Problems

Problem solving, as opposed to verbal problem solving, is broadly defined to include the following tasks: finding an answer to a question, finding a new way of looking at familiar things, collecting and interpreting data, finding a pattern or relationship. A "good" problem has one or more of the following characteristics:

It is open-ended. The problem can be interpreted in different ways, several procedures for arriving at the solution are possible, or more than one solution may exist.

It provides for maximum involvement on the part of the pupils and minimum teacher direction.

It leads to further problems.

It can be integrated into other areas of study.

The problems used with the prospective teachers could be classified into three categories. Several examples of each category are listed.

1. *Questions about ourselves*

What is our favorite TV show (or actor)?
In what month do most of our birthdays occur?
What did (do) we eat for breakfast?
Describe a typical grade___student (teacher).

2. *Experiments and investigations*

How does a thumbtack land?
Say something about the frequency of the vowels on a page of your reader.

Roll a die many times and say something about the outcome.

Roll two dice many times and say something about the sum (difference, product).

What is the largest open box that can be constructed from a sheet of cardboard 7 cm by 10 cm?

3. *Measurement problems*

Compare pulse rate "before" and "after" exercise.
Compare wrist, neck, and waist measurements.
Are you a square? Is anyone? Compare height and arm span.
Predict a person's weight in kilograms from his or her height in centimeters.

The Settings

The problems are presented to the students in a variety of settings. After the problems are solved, the advantages and disadvantages of each setting are discussed. It is anticipated that when the students get classrooms of their own they will try these settings or, as a result of having been exposed to them, discover some settings that best suit their personality characteristics or preferred teaching styles. The following settings were included:

• Each student works on his or her own to solve the problem. (The students are encouraged or challenged to think of interesting or different ways of presenting the solution to the problem.)

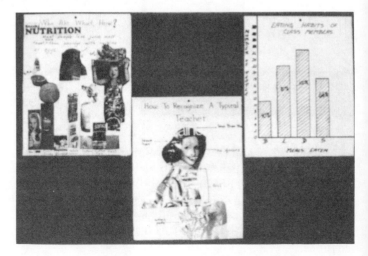

Both Werner Liedtke and James Vance are associate professors at the University of Victoria, British Columbia. They teach preservice and in-service mathematics education courses, conduct seminars, and supervise student teaching.

- Students work together in groups of two or more to solve the problem.

- Four or five different problems are presented, with each of four or five groups of students working on a different problem. A rotational system is set up to insure that each group works all problems.

Other possible settings are discussed and related to possible classroom settings. Two of them follow:

- Week's plan: Four problems and one teacher-directed lesson are prepared. The pupils are separated into five groups. Four of the groups solve the problems, while a lesson is taught to the fifth group. The pupils rotate after every session.

- Problems are placed in an activity corner and each pupil signs a contract to solve a problem in a given period of time, or the pupils are asked to choose one of the problems each week.

The Instructions

Regardless of the problem(s) or instructional setting used, the instructions for the students are basically as follows:

- Collect data.
- Organize the data.
- Communicate the results by drawing a graph.
- Include at least one summary statement of the most important result.
- Make up a title.
- Prepare questions for your representation.

In preparing their results, students use magazines to provide pictures that illustrate or support the title, graph, and/or concluding statement.

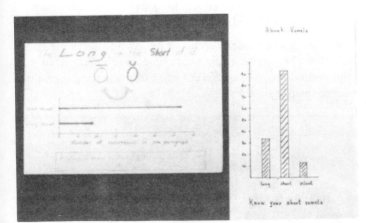

Making use of and developing graphing skills is one of the main objectives of such an activity setting. Students choose different types of graphs for a given situation, and skills of drawing and reading graphs for a given situation are explored. Disadvantages and advantages of certain types of graphs are often discovered during the follow-up discussions.

This problem-solving procedure can be used with young children since graphing can be taught in the very early grades. For example, it is easy for young children to find the answer to the problem, Which pet do we like best? by lining up in front of pictures of pets that have been displayed on the blackboard. Or, instead of children, blocks can be placed in front of the appropriate pictures. For easy comparisons, the blocks are stacked. This representation is then transferred onto graph paper by shading in one square for every block. A title is made up, the parts of the graph are labeled, and one statement is made to summarize the findings.

The Results and Follow-up

Up to a point, the preparation of the solution to the problems is the same for everyone or every group and includes a graph, a title, a concluding statement, and supporting pictures. However, before each group reports its findings, the members of the group are required to make up questions that the audience should be able to answer by looking at the graph. Questions should be of a general nature, and some should involve the use of specific skills. One interesting follow-up task consists of having the students classify a group of questions as to whether or not these can be answered from the data shown on the graph. The questions are either recorded on a chalkboard or bulletin board, or copies are duplicated for each student. The group that makes the presentation or report is then responsible for evaluating the responses and for giving the appropriate feedback.

In settings where individuals work on their own, the results are displayed and similarities and differences are discussed. For example, for the problem in which everyone rolled a die and made a graph about the numbers that turned "face-up" after every roll, similarities and differences between these graphs are discussed. The students are asked to make a guess or prediction, based on their graphs, about a graph that combines all of the data.

A few examples of follow-up questions and activities for some of the other problems are included here:

• Is what you found out about the vowels on one page of your reader true for the reader you used last year? Is it true for the newspaper read by adults?

• Is what you found out for one kind of thumbtack true for another type or make? Why or why not?

• Does the relationship discovered for the wrist, neck, and waist measurements hold for older or younger students?

• Send a letter reporting the information gathered on favorite television shows and actors to the local television station. (One grade-four class that did this received not only an acknowledging letter, but an autographed photograph of their favorite television actor.)

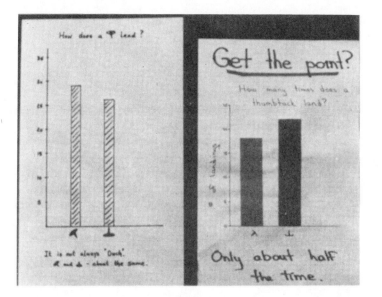

A great variety of strategies becomes evident when students or groups of students are asked to solve an open-ended problem. Comparing these strategies can be an invaluable learning experience. For example, in one session some students were instructed to work on the following problem, either individually or in small groups. The results were to be displayed in some form.

A box without a lid is to be constructed from a sheet of cardboard 7 cm by 10 cm by cutting out equal squares from the four corners and folding up the sides. What is the largest box that can be made?

The first task was to decide how to interpret the question, specifically the meaning of *largest*. Some students decided to find the tallest box, others the box with the greatest surface area. Most of the students reasoned that since boxes are for putting things in, the largest box would be the one that would hold the most (have the greatest volume).

The following are some problem-solving strategies employed by the various students:

Building a model. Many students used paper to construct one or two boxes in order to gain an understanding of the variables in the problem.

Drawing a diagram. The sketch in figure 1 was drawn by most groups at one stage or another while solving the problem.

Fig. 1

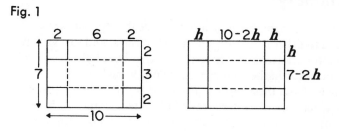

Making a table. (See fig. 2.)

Fig. 2

Height (cm)	Length (cm)	Width (cm)	Volume (cm³)
5	9	6	27
1	8	5	40
1.5	7	4	42
2	6	3	36
3	4	1	12

Drawing a graph. (See fig. 3.)

Fig. 3

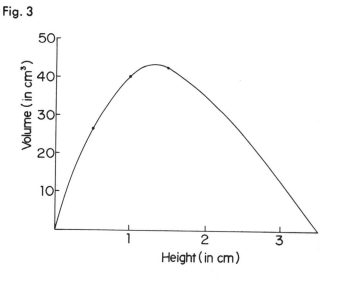

Using successive approximations. From the table and graph, it appears that the maximum volume occurs when the height is somewhere between 1 and 1.5 cm. Using successive approximations, some students determined the height to be between 1.3 and 1.4 cm. A student with a pocket calculator correctly carried the computation to three decimal places, 1.352 cm.

Solving an equation. Many students determined that the formula for the volume of the box is $V = (10 - 2h)(7 - 2h)h$, where h is the height of the box. Two different students who had taken calculus used this tool to solve the problem directly; they set the derivative of the polynomial equal to zero and solved the resulting quadratic formula. One group of enterprising students prepared their solution of the problem on a sheet of paper, then folded the paper to form an open box and presented it to the instructor.

Summary

In their mathematics education course the students are exposed to other activities, simulated settings, and to a variety of manipulative aids. The activities that have been described are included in the course because it is hoped that there is transfer from these settings to the classrooms these prospective teachers will be in charge of in years to come. Perhaps the students will be convinced that an activity or problem-solving setting can be created or set up without expense and without an overabundance of manipulative materials. In addition, it is hoped that the students will realize the value of graphing as a skill and strategy for finding the answer to a question or solving problems.

Reference

Biggs, E. E., and J. R. MacLean. *Freedom to Learn—An Active Approach to Mathematics.* Don Mills: Addison-Wesley (Canada) Ltd., 1969.□

Problem Solving for Teachers

By **Mordecai Zur** *and* **Fredrick L. Silverman**

Many teacher training programs in mathematics education have a weakness in common, namely, concentration on skills at the expense of thinking. This is true for both facets of the preservice education of mathematics teachers, for mathematical content and for pedagogy.

Cognitive mathematical behavior can be classified in three broad categories or levels. The first involves memorization of facts, definitions,

Mordecai Zur is Dean of Education at Bet Berl Teachers College in Israel. At the time this article was written, he was a visiting professor of education at the University of Houston, on sabbatical leave from his work in Israel. Fredrick Silverman is an assistant professor in the College of Education at Louisiana State University in Shreveport. He was a teaching fellow in mathematics education at the University of Houston when the article was written.

rules, and procedures. At this level a student is judged successful if he remembers what he was taught by books or by his teachers and can meaningfully reproduce it in the same context. For example, a student who has been shown a proof that the sum of the measures of the angles of a triangle is 180° can recall, reproduce, and understand the proof, but he does not have the ability to use this information to find the sum of the measures of the angles of a quadrilateral.

The mental activity of generalizing or transferring knowledge from one learning context to another similar one is the second level of mathematical cognitive behavior. For example, the person who operates at this level of thinking should understand the operational algorithms of base ten and also be able to understand these algorithms in some other base, particularly when

he has access to the appropriate basic facts.

The third level of mathematical thinking involves reorganizing and restructuring the variables in a problem so that new relationships, which facilitate the finding of a solution, emerge. It is this "open search" (Avital and Shettleworth 1968) level of mathematical thinking and experience that is missing from so many teacher education programs for prospective teachers of mathematics.

Challenging Mathematical Thinking

Consider the following problem to illustrate the "open search" approach for challenging preservice teachers.

The general setting is that there are three black hats and three red hats. Three persons are blindfolded and a hat is placed on each person's head. When the blindfolds are removed, each person's task is to determine the color of hat on his own head. If a person sees at least one red hat, he must raise his hand. The contestant who first determines conclusively the color of his hat is the winner.

Suppose that a red hat is placed on each contestant's head and the three black hats are hidden. When the blindfolds are removed, the competition begins. Each person sees at least one red hat and raises his hand. Presently one person solves the problem using logical reasoning. What is his line of thinking?

In the open-search approach, reorganizing the problem might include reducing the number of variables, either hats, people, or both. Suppose—searching for a somewhat easier problem—there are two red hats and one black hat distributed among the contestants. This is a simple change in which each person still raises his hand to signify that he sees at least one red hat. Now, for one of the red-hatted persons, either A or B, the situation is easier to interpret. If A sees one black hat on C, he can conclude that B, who also has a red hat, must have raised his hand in the air only because B sees a red hat on A's head. Thus A knows immediately that the color of his own hat is red.

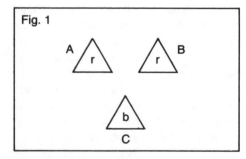

Fig. 1

From the original problem, this is a very natural change. Each person knows either his hat is red or else it is black. Suppose C speculates, for example, that his hat is black. In other words, C conjures up the rearrangment situation just described. He knows that one of his opponents should quickly shout out the answer and give the obvious justification. Because one of C's opponents, in fact, does not have the

solution immediately, C reasons that the only possible cause lies in his speculation on the color of his own hat. Instead of being black, C knows his hat must be red. Thus in this instance, a simple rearrangement of the original problem has made its solution emerge more easily.

What we are suggesting is that problems of this nature be posed for preservice teachers. At first some problem-solving techniques may be modeled. Later, other problems may be posed for the preservice teachers to solve by an open-search technique. In reality, we are suggesting even more than this. What is lacking in so many teacher education courses for preservice mathematics teachers, particularly those at the elementary level, is the spirit of problem posing as well as the activity of problem solving. It is the lack of probing and playing with problems, expanding them beyond their mere statement, that is absent in so many instances. Although someone may charge that puzzle problems of the type just described are somewhat contrived, it must nevertheless be admitted that the problem-solving approach demonstrated here is indeed quite mathematically authentic.

Polya (1945) comes to mind here; first, for his observation that "the first rule of teaching is to know what you are supposed to teach" and second, because a "teacher wishing to impart the right attitude of mind toward problems to his students should have acquired that attitude himself." Helping students learn problem solving is an important goal of school mathematics and it has been widely advocated. What we are saying here is that preservice teachers must engage in problem solving, not simply study mathematical or pedagogical content. The attitude of mind that moves beyond a

problem's boundaries is crucial. To illustrate this kind of elaboration, we are suggesting using problems such as the one of finding the color of the hat.

Extending the problem

After students have successfully solved the first problem, either through an unguided discovery method or through a guided one, the teacher can continue the search through questions.

Is it possible to change another aspect (variable) of the original situation to come up with a different problem?

With three red hats available, one possibility is to change the number of red hats distributed, but still adhere to the agreement that anyone seeing a red hat must raise his hand. The distributions of the hats and the solvability of each case are shown in table 1. So now the original problem becomes one case of many cases in which the solutions are obtained in similar fashion.

Furthermore, students can explore whether increasing or reducing the total number of red and black hats available for distribution has any effect on the solutions of the different arrangements described. For example, suppose there are only two red hats available and only one is used in the contest. Students may be encouraged to explore this situation.

Are the hand cues essential to the solution of the problem?

Here we return to the original problem and examine another change, namely, that of the sign language, raising the hand when a red hat is seen. Students will have to convince themselves that without the hand signals there is no solution to the original problem. In the absence of hand cues, one can solve the problem if the total number of avail-

Table 1

Three red hats and three black hats available

Number of red hats distributed	Distribution			Solvability
	A	B	C	
0	b	b	b	solution immediate, all can solve
1	r	b	b	solution immediate, all can solve
2	r	r	b	solution immediate, A or B solves
3	r	r	r	solution not immediate, all can solve (original problem)

PREPARING ELEMENTARY SCHOOL MATHEMATICS TEACHERS

able red hats is two. The result is shown in table 2.

If the total number of available red hats is one, the case becomes very simple and is left for the reader's exploration. Meanwhile, we proceed to delve further into the problem. Other changes in the original problem may not be so obvious, therefore we point out another feature of the problem, namely, that every participant sees two other participants.

In what way can the number of people seen by each contestant change?

The number of people that a contestant sees can be changed if A, B, and C stand one behind the other in line. (Fig. 2) Then A sees nobody, B sees A, and C sees A and B. With this elaboration, two questions follow: Can the problem be solved? If so, by whom? As before, the answers to these questions depend on the number of red hats available and on their distribution.

Fig. 2

A sees no one, B sees A, and C sees A and B.

One might enjoy determining why some of the solutions are immediate and others are not. The solutions in table 3 all rest on knowing that just one red hat is available. The thinking processes used for the case of one available red hat may be applied to the situations involving two available red hats. For example, here is a sample of thinking for solving the first case in table 4 (two red hats available, each person with a black hat on his head, and the three persons in a line).

Contestant A thinks as follows: If my hat is red, B and C will see it. B will suppose that if his hat is also red, then C would solve the problem instantaneously because there are no red hats left. But since C is quiet, B would know his hat must be black. However, B and C are silent, so A knows his hat cannot be red. Therefore it must be black.

In general, the thinking can be described in the following way. C is the only contestant who could possibly see two red hats; if he did, C would know that his hat was black. Since C does not see two red hats in these examples, he cannot solve the problem. Furthermore, B knows that C is seeing at most one red hat. If A had a red hat, B would see it and know immediately that his own hat was black. Since in these examples B does not see a red hat, he cannot solve the problem. A knows also, from C's silence, that at most there is one red hat between himself and B. If A had a red hat, B would solve the problem. Because B does not solve the problem, A knows his own hat must be black. Thus the cases in table 5, in which A is wearing a red hat, are solved as described. You can see that the reasoning is based on the fact that the last person is the one who can see two hats. The further expansion to the availability of three red hats leads to problems that have no solutions because no one can see three hats. Play with this assertion and see!

Summary

It is important to remember that problems that demand mathematical thinking need not be mathematically complicated. Further, such problems should be relatively easy to extend beyond their bounds so that preservice teachers can probe them and play with them to acquire and open-search frame of mind.

Table 2

Two red hats available and no hand signals

Number of red hats distributed	A	B	C	Solvability
0	b	b	b	not immediate, all can solve
1	r	b	b	not immediate, B or C solves
2	r	r	b	immediate solution, C solves

Table 3

One red hat and three black hats available; contestants in a line, one behind the other

Number of red hats distributed	A	B	C	Solvability
0	b	b	b	not immediate, A or B solves
1	r	b	b	immediate, B or C solves
1	b	r	b	immediate, C solves
1	b	b	r	not immediate, A or B solves

Table 4

Two red hats and three black hats available; contestants in a line, one behind the other.

Number of red hats distributed	A	B	C	Solvability
0	b	b	b	not immediate, A solves
1	b	r	b	not immediate, A solves
2	b	r	r	not immediate, A solves
1	b	b	r	not immediate, A solves

Table 5

Two red hats and three black hats available; contestants in a line, one behind the other

Number of red hats distributed	A	B	C	Solvability
2	r	r	b	immediate, C solves
2	r	b	r	not immediate, B solves
1	r	b	b	not immediate, B solves

References

Avital, Shmuel M., and Sara J. Shettleworth. *Objectives for Mathematics Learning: Some Ideas for the Teacher.* Bulletin No. 3. Toronto: The Ontario Institute for Studies in Education, 1968.

Polya, George. *How to Solve It.* Princeton, New Jersey: Princeton University Press, 1945.

Teaching Problem Solving to Preservice Teachers

By **Stephen Krulik** *and* **Jesse A. Rudnick**

Problem solving! Problem solving! Problem solving! Wherever mathematics teachers turn today, they are confronted by these two words, whether it be when reading NCTM's *An Agenda for Action,* current educational sections of their newspapers, professional journals, or when attending a mathematics education conference. In fact, at the annual NCTM meeting in St. Louis approximately 20 percent of the sections had *problem solving* in either the description or the title of the talk.

We are pleased that this vital topic is receiving the attention it deserves. This special edition of the *Arithmetic Teacher,* for example, illustrates the Council's commitment to emphasis on problem solving in the mathemathics curriculum of the 1980s.

If we are to develop the problem-solving capabilities of children, it is necessary to prepare their teachers in this crucial art and skill. The preservice teacher preparation program is a good starting point. The preservice teacher is young, full of enthusiasm, and has not as yet adopted the erroneous belief that teaching specific algorithms to solve specific problems is teaching problem solving.

Stephen Krulik and Jesse Rudnick are professors at Temple University in Philadelphia, where they conduct courses in mathematics education and microcomputers at both the undergraduate and graduate levels. Krulik was editor of the 1980 NCTM yearbook on problem solving and Rudnick was a member of the editorial panel. Rudnick is also a former NCTM director.

If a teacher is to be an effective guide for the learning of problem-solving skills, then he or she must first become a problem solver. This is not necessarily the case in other professions. For example, coaches instruct their athletes in the performance of tasks that they themselves may not be able to perform. Architects, although they design a building, are not capable of constructing it. It is impossible, however, to teach problem solving if the teachers themselves are not adequate problem solvers.

Fundamental to a preservice program for teachers of mathematics is the clarification of the meaning of the terms *problem* and *problem solving.* Many teachers use the words *question, exercise,* and *problem* interchangeably, but there are differences. As an example, let's look at the following:

What is the product of 7 and 6? or What is 7 × 6?

Is this a question? An exercise? A problem? Or is it all three? It depends! If we ask this of our future mathematics teachers, we expect an immediate response. The answer, 42, should be automatic—a recall based on many years of experience with multiplication. For this audience, What is 7 × 6? is a *question*—it involves only *recall.*

On the other hand, if we ask this of an elementary school student who has been studying the basic multiplication facts, then the purpose of our inquiry is quite different. Here, we are providing drill and practice for that youngster. Thus "What is 7 × 6?" is now an *exercise*—it involves *practice.*

Finally, we ask this of a first or second grader who has not as yet learned the meaning of multiplication. Now the meaning of *product* must be discussed and the youngster led to understand that 7 × 6 means taking 7 groups of 6 objects and finding the total number of objects. To this students we have presented a *problem*—something that requires *thought.*

Thus a question requires recall, an exercise provides drill and practice, and a problem requires careful thought and synthesis of knowledge. *What is a problem at one time to one person may be an exercise or even a question at some later stage of that person's mathematical development. Furthermore, what is a problem for one person can easily be an exercise or a question for someone else.*

A preservice program should also make it quite clear to the students that problem solving is a process. As a process, problem solving is the means by which an individual uses previously acquired knowledge and understanding to satisfy the demands of an unfamiliar situation. The student must synthesize what he or she has learned and apply it to the new and different situation. We have previously stated that what is a question for one person can easily be an exercise or even a problem for someone else, or to the same person at another time. Yet a creative teacher can take any question, any exercise, or any routine word problem and use it as a basis for a worthwhile problem-solving experience. It all depends on the way in which the teacher presents the material.

We do feel it is important that teachers accept the notion that the goal of teaching problem solving is to have students successfully develop a process; that although the answer is important, it is the process that is at the heart of problem solving. In the real world the final answer is the important thing, but our responsibility as teachers is to help students develop a process for finding the answer.

Becoming a Problem Solver

A preservice program should take future teachers through the same kinds of activities and experiences that they will require of their own students a few years hence. Like any other skill, problem solving requires lots and lots of practice.

Figure 1 shows a flowchart of the problem-solving process. The boldly outlined command boxes indicate the procedures that are basic to the problem-solving process. In our classrooms this set of rules occupies a central position. Reference is constantly made to it in our attempt to help the preservice teachers become good problem solvers. The same flowchart will later serve as the basis for their own teaching.

Select a strategy

For most preservice teachers, the most difficult and important part of the problem-solving approach is to "select a strategy." A wide variety of problem-solving strategies must be identified and illustrated by the use of many problem situations. Some problems and solution strategies just naturally fit together. Indeed, more than one strategy may be necessary to solve a particular problem. Preservice teachers should be encouraged to find multiple solution strategies (fig. 2). *It is important to continually emphasize that the goal of problem solving lies in the solution process, not necessarily in the answer itself.*

To illustrate the application of more than one solution strategy, consider the following problem:

There are 16 football teams in the Continental Football League. To

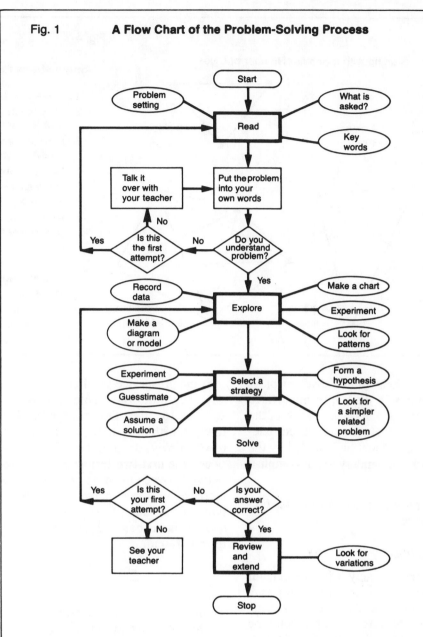

Fig. 1   **A Flow Chart of the Problem-Solving Process**

Adapted from Krulik, Stephen and Rudnick, Jesse *Problem Solving: a Handbook for teachers* Allyn and Bacon, Inc. Boston, Mass. 1980

Fig. 2

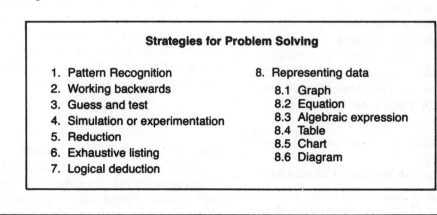

Strategies for Problem Solving

1. Pattern Recognition
2. Working backwards
3. Guess and test
4. Simulation or experimentation
5. Reduction
6. Exhaustive listing
7. Logical deduction
8. Representing data
    8.1 Graph
    8.2 Equation
    8.3 Algebraic expression
    8.4 Table
    8.5 Chart
    8.6 Diagram

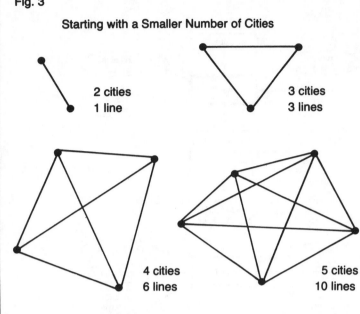

Fig. 3

Starting with a Smaller Number of Cities

2 cities
1 line

3 cities
3 lines

4 cities
6 lines

5 cities
10 lines

Fig. 4

Suggestions for Teaching Problem Solving

1. Create an atmosphere of success.
2. Provide an ample supply of challenging problems for problem-solving practice.
3. Help students to develop techniques to read problems analytically.
4. Require students to create their own problems.
5. Have students work together in pairs or small groups.
6. Encourage students to attempt alternate strategies.
7. Raise creative, constructive questions when leading a problem-solving discussion, as a model for students to emulate.
8. Require students to estimate their answers.
9. Use strategy games to develop the problem-solving process.
10. Have students make a flowchart of their own problem-solving procedures.

conduct their annual draft of players, teams in each city must have direct telephone lines to each of the other cities. How many direct telephone lines must be installed by the telephone company to accomplish this?

Three possible approaches are presented here.

First strategy

Some students may wish to act out this problem. For instance, they may set up 16 desks and tie pieces of string from one desk to another, counting them as they go (experimentation).

Second strategy

Others may examine the problem starting with a smaller number of cities (reduction), keeping track of their data (organized listing), and looking for some pattern that appears (pattern recognition). Using this approach (fig. 3), we begin with two cities, then consider three cities, then four cities, and so on. The data collected can be recorded in a table (see table 1).

Table 1
Data on the Number of Direct Lines

Number of cities	1	2	3	4	5	…	16
Number of lines	0	1	3	6	10	…	?

Some preservice teachers will actually continue the table for 6, 7, 8, . . . , 16 cities. Others may be able to work with the pattern in the second row, namely, an increase of one between the first two terms; an increase of two between the next two terms; an increase of three between the next two terms, and so on, until the sequence is recognized or the general term is found.

$$\frac{n(n-1)}{2}$$

Students with a more sophisticated mathematics background may recognize this as the method of finite differences.

Third strategy

Other preservice teachers may use *logical deduction*. Since each city is connected to every other city, there will be 16 times 15 connections; but since *A* to *B* is the same as *B* to *A*, there will be one-half as many connections needed, resulting in

$$\frac{(16)(15)}{2}$$

Thus, 120 telephone lines needed.

Some people may recognize this problem as another formulation of the well-known "handshake" problem, and may know how to solve it right

away. For these people, this is not a "problem," but merely a "question."

Notice that this problem was solved in several ways with a variety of strategies. Experience shows that no one strategy is any more important or valuable than any other. Pattern recognition, however, seems to be called upon in conjunction with many of the other strategies.

We have illustrated briefly a technique that we use in our preservice program to help our students become problem solvers. Now let us look at some suggestions that will help these preservice teachers teach problem solving to their students.

Teaching Problem Solving

We still believe that the best method for teaching problem solving is for the teacher to be a good model, but there are specific things that teachers should do in the classroom. Ten suggestions for teaching problem solving are listed in figure 4. Let's take a closer look at "helping students to develop techniques to read problems analytically."

Being able to read the problem analytically is crucial to successful problem solving. After all, if students cannot interpret what is being asked or what information is supplied, then

PREPARING ELEMENTARY SCHOOL MATHEMATICS TEACHERS

they can hardly be expected to solve the problem. Many teachers assume that if students can read the words, they understand the context and can interpret the information. This is far from true.

What steps can classroom teachers take to help students in this vital area? Five are essential:

1. Have students restate the problem in their own words and discuss with the teacher and with the class what is taking place.

    a. Can they "see" what is happening?

    b. Have they discovered all of the "given"?

    c. Have they been able to eliminate the extraneous data?

    d. Are they aware of the hidden data, such as different time zones, units of measure, and so on?

2. Discuss the meanings of words.

    a. Mathematical terms that might be completely unknown to some students.

    b. Words with multiple meanings. (Many words assume a meaning in a mathematical context that is different from everday usage; for instance, *prime*, *count*, *root*, *volume*, *chord*, and so on.

3. Have the students provide appropriate questions for stories that you present.

4. Have the students provide stories for questions that you present.

5. Provide problems for which the student must determine whether there is an excess of, deficiency in, or just enough information. In the first two cases, they should determine what is excess or supply what is deficient.

With regard to number two in figure 4, providing an ample supply of challenging problems for problem solving, we suggest the following:

Get a supply of 6-by-9 file cards in a variety of colors, on which you will write the problems you collect. Decide upon your own color coding system. (For example, white cards might contain the simplest problems, blue the next simplest, and so on, with green cards having the most difficult problems. When you find or create problems that you feel would be good vehicles for problem-solving experiences for your students, put them on these cards. On the backs of the cards you can list the strategy or strategies that the problem suggests.

Problems can be found in a multitude of places. The *Mathematics Teacher* and the *Arithmetic Teacher* are valuable sources, as are other journals. The IDEAS section of the AT is another excellent source, as are conferences at which talks on problem solving are given. Finally, go "back to basics"—very far back! Examine old mathematics text books, some over 100 years old. Many of the problems in these books are excellent vehicles for problem solving.

Whatever the source of the problems, in all cases preservice teachers should have an opportunity to actually solve these problems and to identify the strategy or strategies they used in the process. This not only provides them with new problems for their own problem file, but also gives them additional practice in becoming problem solvers.

In this brief paper we have described what can be done to help preservice teachers become good classroom teachers of problem solving. ◆

Teacher Education

Elementary Teacher Education Focus: Problem Solving

By **John F. LeBlanc**
Indiana University, Bloomington, IN 47405

Somebody in a gathering or at a party is often likely to announce to the group, "I've got a problem for someone here to solve! Who would like to hear it?"

The reaction of the group is almost dichotomous. Some of the people groan and return to what they were doing before the interruption. These individuals have no interest in hearing or solving the problem.

The other portion of the group quickly and eagerly asks, "What's the problem? I can solve it." These individuals cannot wait to try to solve a problem or puzzle.

Preservice teachers' (PSTs') and practicing teachers' reactions to problem solving are similar to those of the general population. The attitude of one group of PSTs and of teachers is favorable toward problem solving. They seem to have the following characteristics:

- They enjoy the challenge of a problem or a puzzle.
- They have some confidence that they will be able to understand and solve the problem.
- They almost "beam" at the prospect of this intellectual exercise.
- They are anxious to be the first to solve the problem and explain it to others.
- They are anxious to improve on another's solution if they are not the first to solve the problem.
- They feel free to select a variety of processes or methods in their search for a solution.
- They do not feel restricted to the use of *formal* mathematical approaches, such as learned algorithms or formulas.

The attitude of another group of PSTs and teachers is quite the opposite. This group does not like mathematical problem solving.

They seem to have these characteristics:

- They detest solving problems.
- They know they cannot solve problems.
- They have no confidence in themselves when faced with, or in attacking, a problem or a puzzle.
- They are bored and quite "turned off" by the enthusiasm of others toward solving problems.
- If the problem is "forced" upon them, they protest that they neither understand the problem nor have any idea how to get started.
- If they do try to solve the problem, they feel they should use previously learned formal techniques involving standard algorithms or formulas.
- They have no interest in hearing the solution or in discussing assorted processes of solution.

Perhaps the fact that the population of prospective and practicing teachers has such dichotomous sets of characteristics related to involvement in problem solving has been exaggerated. Unfortunately, however, a set of teachers and would-be teachers exists whose feelings toward problem solving are similar to those of the latter group.

How can these attitudes toward problem solving be changed?

Are these attitudes based on reality, that is, are these people unable to solve problems?

These two questions seem to be related. A population of PSTs and teachers seem not to have had success in solving problems. Their attitudes toward problem solving will probably not be changed until they experience some success in solving problems.

(It does not seem necessary to ask this question here, but what kind of influences do teachers have on pupils' attitudes toward problem solving if the teachers' attitudes and experiences toward it are so unhappy?)

Not surprisingly, attitude and performance go hand in hand. Teachers' attitudes toward problem solving will improve as their performance in solving problems improves. The statement that "one has to be a problem solver to teach problem solving" may or may not be true. But a teacher who has had some successful experiences in solving problems is more likely both to convey a more favorable attitude on this topic and to import a sense of enthusiastic confidence to pupils than a teacher who has not had such experiences.

The question for teacher educators then becomes, *"How can we, as trainers of teachers, provide successful problem-solving experiences for teachers?"*

Two methods can be used to offer problem-solving experience and to teach problem solving to teachers—informal and formal. Both of these methods, by the way, are methods that should be modeled in teacher-education courses, to be imitated later by teachers with pupils.

PREPARING ELEMENTARY SCHOOL MATHEMATICS TEACHERS

The informal method is based on a widely held belief among mathematics educators that the best way to become a problem solver or to teach problem solving is to solve problems. The teacher educator should continually supply the PST or in-service teacher with problems to solve. Accompanying these problems should be appropriate encouragement, motivation, and recognition. Informal problem-solving instruction can be effected in several ways, two of which are described here.

One way is to assign problems throughout the course. At the beginning of the term in a content, methods, or in-service course, the teacher educator can present a nonroutine problem at the end of each class period to be solved by each class member. The students should be allowed to elicit solutions from their roommates, family members, and so on, *as long as* they understand and can explain the solutions. Students whose confidence and success in problem solving has been lacking or very limited will need some help. This help can come, in part, from hearing how others attack and solve the problem.

In a succeeding class session, have a student present a solution. Follow this presentation with a brief discussion of other solutions. Another problem should then be assigned. One should make sure that each class member is required to present a solution at least during once during the term.

An alternative method of providing problems is to put one on a bulletin board each day or week or to hand a dittoed copy to each member of the class at each session. Encourage the participants to use this informal technique as teachers. This informal instruction in problem solving should begin at the first session of the term and continue throughout the course.

A second way to effect informal instruction in problem solving is to encourage participants to try to adopt a problem-solving attitude toward every aspect of learning or teaching. For example, in a session in which participants are given the assignment of correcting and analyzing (diagnosing errors) some pupils' papers, the participants should be challenged to identify not only the errors but also the possible causes of the errors, as well as to suggest some solutions for remediation. They should see such diagnostic efforts as problem solving.

PSTs and teachers should be encouraged to develop a generalized problem-solving attitude toward all learning and teaching, so that they will be ready to accept and even encourage such ways of thinking in their teaching. Many routine algorithms can be viewed as problems by young pupils who have not been previously taught the formal algorithms. For example, when presented with an addition problem involving regrouping (not previously taught), a young pupil might solve it using counting by tens or other techniques that are not standard. The teacher with a generalized problem-solving attitude will encourage such thinking on the part of the pupils.

Formal instruction in problem solving should take place as a unit in a concentrated period of time. This formal instruction should be a part of every program of education for pre-service and in-service teachers. An instructional sequence that has been successful for me is outlined in the following paragraphs. This suggested sequence has three steps.

1. In the first step students should be given problems to solve. These problems should be a combination of both routine (readily solved using a stardard algorithm) and nonroutine (student doesn't have an algorithm readily available for solution) problems. Both types of problems are important, since the participants should be encouraged to solve even routine problems using nonroutine techniques. The participants should be encouraged to try to solve them on their own—but after solving a problem or at least *trying* to solve it, they should ask others how they might solve it.

As a culmination of this problem-solving step, students should report on how they or others solved the problem. The emphasis in the reporting period should be a discussion of the various techniques used. The answer, which most students have been led to believe is the most important aspect, should be downplayed. The techniques or processes should be identified and labeled. Some students will favor one technique for one problem and a different technique for another.

This discussion stage is critical for the students' growth in confidence. Peer teaching and listening can become powerful tools in the students' growth in problem-solving confidence. An atmosphere of acceptance of each and every participant's thinking and efforts is critical. The teacher educator must be sensitive to the lack of confidence that many participants will exhibit. A gradual building of confidence should be an important concern—and outcome—of these discussion sessions.

The second step in the formal instructional sequence is to identify a problem-solving model that can be used by prospective teachers both as *their* model for solving problems and for their use in teaching children later. Polya's model of *understanding* the problem, *planning* to solve it, *solving* it, and *evaluating* the answer is excellent. Prospective teachers can be aided in their own problem-solving abilities and later in teaching problem solving by some analysis of each aspect of this model.

As specific problems are presented, the problem solvers should be encouraged to help their *understanding* of the problem by thinking about—

- the meanings of the symbols;
- the meanings of the words;
- what conditions are inherent in the problem;
- what facts are given; and
- what is "asked for" in solving it.

Participants can help themselves grow as problem solvers—and use the same techniques as teachers—by asking specific questions related to broadening the understanding of the problem. As these questions are asked of oneself or the pupils, the information given in the problem's statement, as well as what is being sought in its solution, can become clearer.

In *planning* to solve the problem, the prospective teachers should be encouraged to "run a search" through their repertoire of techniques at their disposal for solving a problem. Here drawing a sketch, a diagram, or writing an equation could help to bridge the gap between the problem, as it is understood, and a procedure for its solution. This planning aspect could be thought of as an interpretive aspect, one linking the problem with a potential solution process.

In *solving* the problem the prospective teachers should be encouraged to try *any* technique to get started. If the technique proves fruitless, they have undoubtedly gained more insight into the problem. Another trial using another technique may be more rewarding.

In the final aspect of the model, *evaluating* the solution, prospective teachers should check the solution to see that it makes sense and that any computation performed in arriving at the solution is accurate. As part of this final aspect, a review of the earlier discussion on how others solved the problem can help to broaden the repertoire of solution processes.

The Polya model for problem solving is well accepted among mathematics educators. The presentation and use of this model can be enhanced by focusing on each aspect of the model as suggested. The participants seem to gain confidence in the model as each aspect is reviewed during a discussion of specific problems.

The third and final step in the formal instructional sequence in problem solving is to have each prospective teacher collect a set of problems for later use. A few years ago ready sources of problems and puzzles were hard to find. Now several publications contain problems that are appropriate for developing problem-solving skills. Some care must be taken in unqualified recommendations of these publications. Prospective teachers might be challenged to form groups to review some of these problems for classroom use. This task in itself could become a nice problem-solving assignment.

Prospective teachers reflect the general population's characteristics in their attitudes and performances in problem solving. Some enjoy it, others do not. Prospective teachers' attitudes toward problem solving will become more favorable as they become better problem solvers. A preservice or in-service program of teacher education that emphasizes an informal and a formal focus on problem solving is recommended.

Specific Materials and Techniques

THE articles in this section identify a variety of specific materials and techniques for use by teacher educators.

In "How Teacher Educators Can Use the *Arithmetic Teacher*," O'Daffer elaborates on ways of using the journal in a methods class. Schmalz describes her system for acquainting students with the professional literature in "Assigned Readings from Professional Journals—a Suggested Procedure." Burton advocates "Writing as a Way of Knowing in a Mathematics Education Class," suggesting how to employ four writing-based strategies: free writing, writing in journals, in-class writing, and term papers.

In "How Teacher Educators Can Use Manipulative Materials with Preservice Teachers," Young outlines a variety of activities to introduce preservice teachers to the value of using manipulatives when teaching children. In "Hands On: Help for Teachers," Trueblood also deals with preparing preservice teachers to use manipulative materials in the classroom, discussing the nature, scope, and variety of settings and activities used in an elementary and early childhood teacher education program.

Three articles deal specifically with laboratory settings. Clarkson, in "A Mathematics Laboratory for Prospective Teachers," describes a methods course emphasizing the mathematics laboratory approach and giving an important role to problem solving. Five different types of laboratory or field experiences are discussed in "Preservice Laboratory Experiences for Mathematics Methods Courses," by D'Augustine. "Mathematics Methods in a Laboratory Setting," by Sherard, identifies activities for the teaching of geometry, measurement, probability, and statistics in a mathematics laboratory with an emphasis on active involvement using manipulative materials.

In "Using Microcomputers with Preservice Teachers," Schroeder provides three general considerations and some specific examples to guide work with microcomputers in mathematics education classes. Sadowski discusses the knowledge, skills, and attitudinal goals of a program for preparing teachers to use microcomputers and describes a computer literacy course for preservice teachers in "A Model for Preparing Teachers to Teach with the Microcomputer." In "Computer Training for Elementary School Teachers and Elementary School Computer Specialists," Polis argues for a general course in the use of computers, followed by the integration of applications across the curriculum.

Specific learning experiences to help preservice teachers develop an awareness of the mathematics curriculum and a working philosophy of mathematics instruction are detailed by O'Daffer in "Helping Preservice Teachers Develop an Awareness of Curricular Issues."

In "Mathematics Educators: Establishing Working Relationships with Schools," Trafton discusses the importance of close contact and interaction with schools and offers suggestions about how such relationships can be developed.

How Teacher Educators can use the Arithmetic Teacher

By **Phares G. O'Daffer**

Would you want more information if you received a letter containing this announcement?

Now available. Valuable new teaching resource materials for teacher educators. Try them—you'll like them. Write for a free sample.

Recently a colleague described a workshop for inservice teachers in which she handed out Xeroxed sheets containing some excellent ideas that teachers could use the next day in their classrooms. After the initial excitement about trying these ideas, the teachers were informed that they had come fom the *Arithmetic Teacher.* Several of them were surprised to learn about the journal or to learn that it contained so many useful ideas for teaching mathematics.

My own experiences have convinced me that mathematics educators find it worthwhile to investigate the possibilities the *Arithmetic Teacher* has for providing valuable experiences for our students. It may well be one of those valuable teaching resources that we have essentially overlooked.

For example, how could the *Arithmetic Teacher* help us in our methods classes for preservice teachers? An initial activity at the beginning of the semester might be to have the students do a "scavenger hunt" in the *Arithmetic Teacher.* A predesigned assignment sheet could be prepared. Questions leading students to look at the special features of the *Arithmetic Teacher*—the "Let's Do It" article, the

Phares O'Daffer is a professor of mathematics at Illinois State University, Normal, Illinois, and a member of the Editorial Panel.

"IDEAS" activity sheets, the "From the File" cards, the "Reader's Dialogue" section, the "Reviewing and Viewing" columns, and the "One Point of View" essay—could be asked. This assignment would acquaint the students with the *Arithmetic Teacher.* It could also include questions requesting that the students bring back an important idea found in each of these sections. An experience like this could be the first step in acquainting prospective teachers with a professional journal that is important to them. It can help them develop the habit of reading journals such as this and searching for new ideas that will make their teaching a continuing growth experience.

In my methods class, I follow this initial experience with an assignment requiring the students to read at least ten articles from the *Arithmetic Teacher* during the course. I ask them to make a card file (using 5-by-7-inch cards) which include for each article (1) bibliographic data, (2) summary and critique of major ideas, and (3) specific suggestions for use in the student's classroom. At the end of the course, students invariably indicate that this reading assignment has been enjoyable and valuable for them. I believe that several differences I have noticed on pre- and post-attitude inventories have come from the broadening effect resulting from these readings.

Another technique that has been suggested is that the methods instructor make a bibliography card for each article in each new issue of the *Arithmetic Teacher.* The cards could be classified by key areas of current emphasis in mathematics education (problem solving, calculators, computers, gifted

programs, and so on). As these topics come up in the methods course, selected students are assigned to read an article and give an "extra credit" report, make a bulletin board, or provide a duplicated critical summary of the article for others in the class. Because of the delays in publishing, it is well-known that methods texts are somewhat behind in reflecting current trends in mathematics education. Because articles in key areas are processed quickly, the *Arithmetic Teacher* is able to provide students with an up-to-date interpretation of current trends.

When persons who are now teaching stop by for a visit to their old methods instructor, the universal admonition for the preservice teachers is, "Have them make files and collect problems and activities now. They won't have time to do it later." The *Arithmetic Teacher* can be a great help in this endeavor. Students can be encouraged to start a card file based on the "From the File" and "Challenge: for Able Students" departments. Suggest, also, that they use ideas from other articles in the journal to make their own file cards to add to their collection. It should also be noted that these particular features in the *Arithmetic Teacher* provide an early opportunity for preservice teachers with ideas to get them published. It is quite a thrill when a preservice teacher submits an idea for "From the File" and sees it in a subsequent issue of the *Arithmetic Teacher.* It is also quite a confidence builder.

Since the "Let's Do It" feature almost always contains suggestions for physical materials which can be

How Teacher Educators can use the Arithmetic Teacher

teacher-made and used in the classroom, a useful assignment for preservice teachers involves actually making materials and preparing a lesson plan for incorporating these materials in a developmental sequence.

We often present ideas to preservice teachers involving pre-book experiences and follow-up experiences. These include motivational, developmental, practice, enrichment, and remediation activities. The "IDEAS" section of the *Arithmetic Teacher* is a rich source for these purposes. Students could be encouraged to make file folders of "IDEAS" worksheets classified according to topic, grade level, or objectives. The "IDEAS" sheets could also be categorized according to the pre-book and follow-up experiences.

As important as the "What do I do on Monday?" ideas are, it is also important to encourage preservice teachers to expand their thoughts on trends, philosophy of teaching mathematics, and pedagogical techniques. One way to accomplish this is to assign readings that include the editorials in "One Point of View." To guide students in their reading or thinking, questions like the following can be asked: "What was the author's major point(s)?" "Do you agree with the position taken? Tell why or why not. Use another reference to support your position."

As students read other articles in the *Arithmetic Teacher*, it is instructive to ask them to rate the article on a scale of 1 to 10 and to make some brief comments with regard to their rating of the article. I have learned a lot about a given preservice teacher's philosophy of education through this particular activity.

I also have found it quite useful to present information regarding membership in the NCTM and subscription to the *Arithmetic Teacher* near the end of the course. By that time students have seen the value of this journal and have developed a habit of reading the articles and using the ideas. I believe the value of this association with a professional organization is significant and can make the difference between a teacher who "stays alive" and one who "dies on the vine."

The *Arithmetic Teacher* is not only a valuable adjunct to the preservice teacher program, but also a useful resource in work with graduate students and inservice teachers. I have talked to several teacher educators who often build workshops around the content of an *Arithmetic Teacher*. The "Let's Do It" section provides material for a make-and-take session. The "IDEAS" and the other articles suggest a variety of techniques for classroom instruction. If teachers are encouraged to share with other teachers both the journal itself and the ideas in the journal, the results could be quite surprising.

Graduate students can be asked to analyze activities and articles in the *Arithmetic Teacher* carefully. Activities could be classified in terms of objectives, learning theories, research findings, and so on, and variations on the activities could be created. Back issues of the *Arithmetic Teacher* could be used to investigate the history of the development of elementary school mathematics and to ascertain how the emphasis has changed over a period of years. This could provide the basis for a discussion of current and future trends in mathematics education. The published results of the National Assessment, reports of classroom research, the editorials, and other more theoretically oriented articles can be used to generate presentations and discussions. The reviews of books and other materials can be compared with the results of analyses made by students using criteria they have developed.

Finally, perhaps the greatest use we as teacher educators can make of the *Arithmetic Teacher* is to read the journal ourselves. It can provide us with an extended background of "ideas that work." We can test these ideas when we work with children, look for variations in other good or related ideas, and pass them on to others. ▼

SPECIFIC MATERIALS AND TECHNIQUES

Assigned Readings from Professional Journals— a Suggested Procedure

By **Sr. Rosemary Schmalz**

In the November 1980 issue of *Arithmetic Teacher*, Lowell Leake describes his efforts and experiences in his class for prospective elementary teachers. Many of my efforts have been similarly directed, but I would like to respond to one point in particular: the assigning of readings from professional journals. Over the past years, I have worked out a system that I have found quite successful in making such assignments.

Our mathematics sequence for elementary teachers is an integration of mathematics and methods consisting of two four-hour semester courses. About six years ago, I began requiring readings from journals, especially from the *Arithmetic Teacher*, as part of the course. My goal was two-fold: to acquaint students with professional journals and to extend their mathematical experiences beyond what we could cover in the classroom.

My first attempt was simply to assign readings and tell the students to summarize the articles. This practice was frustrating to me. As one may guess, the reports were done with varying degrees of care. Being a mathematics teacher, I had little experience in assigning grades in an objective way to written reports, especially when a student handed in a long report that seemed to have missed what I considered the major point of the article. For lack of other options, I would ask students to redo unsatisfactory reports. I was also teaching a

Rosemary Schmalz teaches at the University of Scranton, Scranton, PA 18510. She teaches computer science with emphasis on teaching computing in the liberal arts curriculum.

Saturday class for noncertified teachers at that time. They had felt inconvenienced to have to go to the library to read the articles in the first place. They considered my demand outrageous that they should have to return to the library to redo them.

Out of this situation came an inspiration for a technique that I have used for the past five years. Briefly it is this: when I assign an article to be read, I pose specific tasks to be done as a result of reading the article. Included in this article are some examples of the various categories of assigned readings given in the form that I use when making the lists for my students.

Some articles explain a particular method of computation or teaching a computational skill.

- Bell, Kenneth M., and Donald D. Rucker. "An Algorithm for Reducing Fractions." *Arithmetic Teacher* 21 (April 1974):299–300.

  Use this method to test if the fractions

  $$\frac{39}{52}, \frac{1287}{1331}, \text{ and } \frac{49}{72}$$

  are reducible. If they are, reduce them.

- Batarseh, Gabriel J. "Addition for the Slow Learner." *Arithmetic Teacher* 21 (December 1974):714–15.

  Copy the five steps of this method and the example in figure 1. Then do these problems using the method.

  $$498 + 375$$

  and

  $$1148 + 627.$$

Another type of article describes a teaching device.

- Grove, Julia. "A Pocket Multiplier." *Arithmetic Teacher* 25 (March 1978):54.

  Make a pocket multiplier for one set of facts. Give instructions from the article for using it.

- Ziesche, Shirley S. "Understanding Place Value." *Arithmetic Teacher* 17 (December 1970):683–84.

  Make three expansion cards suitable for the grade level at which you wish to teach. If you are considering the upper grades, you may want your cards to describe decimal fractions.

Some articles are of the "helpful hints" variety.

- Myers, Ann C. "The Learning Disabled Child—Learning the Basic Facts." *Arithmetic Teacher* 25 (December 1977):46–50.

  This article is a "gold mine." Read it thoroughly and take clear notes on the five ideas that you like best.

- Leutzinger, Larry P., and Glenn Nelson. "Let's Do It: Counting with a Purpose." *Arithmetic Teacher* 27 (October 1979):6–9.

  Write down the four basic abilities discussed in the article and why each is important. Then, for each of the four, copy two of the activities suggested for developing the ability.

Still other articles discuss interesting topics in number theory that are suitable for children and thus for their prospective teachers.

- Kennedy, Leonard. *Guiding Children to Mathematical Discovery* (Belmont, Calif.: Wadsworth Publishing Co., 1970, pp. 217–20).

  Finish the 8 × 8 magic square on page 219. What is the magic sum for a 9 × 9 square? Make a 9 × 9 magic square.

- Cox, Anne Mae. "Magic While They Are Young." *Arithmetic Teacher* 21 (March 1974):178–81.

  Find at least four perfect patterns in your 9 × 9 square. Color them as shown in the article.

Similarly, some articles offer enrichment and extensions to topics other than number theory, such as metric and nonmetric geometry.

- Bernstein, Robert, and Alan Barson. "Decoding Student Names, or If *Alan* Is 42, Then *Robyn* Must Be 82." *Arithmetic Teacher* 22 (November 1975):591–92.

  Write your first and last names in the printing style of your choice. Use the key in figure 1 to compute what your first name is worth. What is your last name worth?

- Brougher, Janet Jean. "Discovery Activities with Area and Perimeter." *Arithmetic Teacher* 20 (May 1973):382–85.

  Draw four noncongruent figures each with the same perimeter but with different areas. Draw two noncongruent figures with the same perimeter and area.

Lastly, some articles comment on the more general issues or goals of mathematics education.

- Suydam, Marilyn N. "The Case for a Comprehensive Mathematics Curriculum." *Arithmetic Teacher* 26 (February 1979):10–11.

  The author compares traditional "basic skills" with a more current list of skills. Summarize her arguments in favor of the current list. Give your own reaction to the article.

- Ernest, John. *Mathematics and Sex* (Santa Barbara, Calif.: University of California, 1976, pp. 1–10).

  List five major points demonstrated by the author's research. Comment on how these relate to your own experience.

Every year I review the articles, delete some from the lists, and add new ones—especially current ones. I make a reading assignment about every two weeks on the topics that we are currently covering in class. My assignments used to include ten to twelve articles. I found that many of my students grew to hate the assignments, complaining especially about the *amount* of work involved. During the past two years, I've reduced my lists to five to six articles, which forces me to be very selective. Although I wonder if I have not succumbed to the desires of the less ambitious students in watering down my demands, I do find that my current students react more favorably to the assignments and feel that the readings are a useful part of the course. I salve my conscience by telling myself that although they are doing less work, they are developing a more positive attitude toward professional reading.

Because each task is so well defined, grading the reports objectively is easy for me. I assign a certain number of points (usually between 2 and 8) to each reading. Full or partial credit is given depending on the completeness and correctness of the response to the directions. I can always point out the exact reason that a student has not earned full credit. The raw scores are recorded and then averaged at the end of the semester to get a student's reading grade, which is averaged with the student's test grades. The cumulative final examination is the third component of the overall grade. The grades earned on the readings usually match the test averages fairly well and almost never inflate the grade, probably because careful reading and accurate responses are required to earn full credit for the report. These very characteristics of careful reading and accuracy are lacking in a student of limited ability or ambition.

Making these readings as part of my class has benefited me as well. I am forced to keep abreast of current journals. About 90 percent of my readings are from the *Arithmetic Teacher*, but some are from *School Science and Mathematics*, as well as from various books such as the NCTM yearbooks, methods textbooks, and books on the history of mathematics. I have become very familiar with our library's holdings and have been able to recommend additions to the collection.

Like Leake, I always take time to give each class a good description of the services of NCTM, being sure to point out that they can get their first year of membership for half price if they join before they graduate. These assignments are helping these future teachers to develop an appreciation of professional reading and professional organizations that I hope will carry over into their professional lives.

Reference

Leake, Lowell. "Some Reflections on Teaching Mathematics for Elementary School Teachers." *Arithmetic Teacher* 28 (November 1980):42–44. ◗

Writing as a Way of Knowing in a Mathematics Education Class

By **Grace M. Burton**

We do not think and then write, at least not without putting an unnecessary handicap on ourselves. We find out what we think when we write, and in the process put thinking to work—and increase its possibilities.

Frank Smith (1982)

Please read what is in the box below and follow the directions given.

> Take two clean sheets of paper and a writing instrument. Put the instrument to the paper and write, without stopping, on the topic "Using Writing in a Mathematics Methods Class." Do not try for complete sentences; ignore punctuation and paragraphing; do not reread anything you have written. Please begin now, before you read further. If you can think of nothing to write, write the topic over several times.

Free Writing

What you have just done (if you followed the directions in the box) is called free writing. An example of my free writing on this topic can be seen

Grace Burton uses the techniques described in this article to teach undergraduate and graduate mathematics methods courses at the University of North Carolina at Wilmington, Wilmington, NC 28403. She also serves on a campuswide committee on writing.

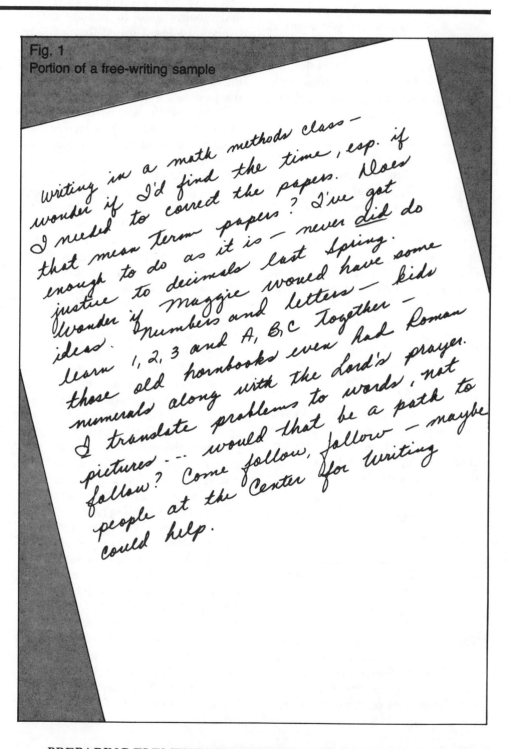

Fig. 1
Portion of a free-writing sample

writing in a math methods class— I wonder if I'd find the time, esp. if I needed to correct the papers. Does that mean term papers? I've got justice to do as it is— never did do justice to decimals last Spring. Wonder if Maggie would have some ideas. Numbers and letters— kids learn 1, 2, 3 and A, B, C together— those old hornbooks even had Roman numerals along with the Lord's prayer. I translate problems to words, not pictures ... would that be a path to follow? Come follow, follow— maybe people at the Center for Writing could help.

in figure 1 Your output won't read just like mine, but perhaps the flavor is similar. Free writing contains disconnected phrases, incomplete sentences, misspelled words, and so on. Unlike much of the composition we have learned, free writing proceeds without a plan. Writers are often amazed that during free writing, the pen begins to trace words almost by itself. This phenomenon, however, is not automatic writing, a popular parlor trick of Victorian times; it is the product of your own mind—perhaps not the conscious, rational left hemisphere but your own mind nevertheless.

Many of us have been taught to believe that every piece of writing we do must conform to all the rules of grammar and syntax—that all the words must be spelled correctly and that all the punctuation marks must be in the right places. Although these restrictions apply to formal, or "pub-

---

Unlike much of the composition we have learned, free writing proceeds without a plan.

---

lic," writing, they are suspended when free writing or other informal, or "private," writing is done. This suspension, once assimilated, can indeed be freeing. Private writing gives an individual a chance to generate

ideas and relationships that can be explored, reshaped, and readied for more formal oral or written treatment.

Many ways exist to incorporate free writing into a mathematics education class. If you were going to lecture on meeting the needs of the learning disabled child, for example, you might ask the class to free write for three or four minutes on learning disabilities as they relate to mathematics. By focusing students' attention on the topic at hand, this activity will allow you to sow the seeds of your lecture on fertile ground. Free writing is also appropriate after a lecture or a film. Before the general discussion, ask everyone to let the words flow on a specific aspect of the presentation. Recording immediate impressions and reactions gives students a chance to gather their thoughts and leads to a more fruitful discussion. Not only will the exchange include more students but it may also take place at a higher cognitive level. If students have difficulty understanding a concept within a lesson, free writing can help them become specific about the areas that are causing a problem. Being able to address the trouble spots directly allows the instructor to plan class time very efficiently.

Journals

Free writing is but one of the powerful writing-based strategies useful in preservice or in-service mathematics education courses. Writing in journals is another technique that costs no money and little time but yields copious benefits in the quality of class time.

Most of us have become accustomed to writing only to communicate with others and may devalue communicating only with ourselves. Journal writing, however, provides a record that can help us discover patterns of thought that produce growth and those that do not. Whereas diaries tend to focus on "what I did," journals extend to "what I thought about what I did."

Like free writing, journal writing is a private, enlightening procedure. Turning off one's internal censor can allow latent ideas to surface. Released

from worries about spelling, capitalization, and sentence structure, my

---

Journal writing provides a record that can help us discover patterns of thought.

---

students often find themselves producing material that surprises even themselves. Often they produce ideas that merit serious, more structured development. Journal writing is a kind of brainstorming with oneself, and the rules of brainstorming apply: Everything must be written down without evaluation, quantity is the goal, and building on expressed ideas is encouraged.

Journals serve as a permanent record of the course for students in ways that ordinary class notes cannot. For students who still have several semesters before their preservice teaching, journals serve as a strong argument for completing each assignment with care. Whether one moves from a behavioristic, a cognitive-developmental, or a humanist perspective, personal reactions to experiences are valuable data and, thus, the writing of journal entries can be appreciated by students regardless of their philosophical persuasion.

Students who are reticient about using a journal can be helped if five to ten minutes of each class is devoted to the procedure and if you act as a model during this period. For both the instructor and the students, immediately recording the contents of, and personal reactions to, a particularly arresting or difficult lesson can be worth several hours of study or reflection at a later time. When the notes are reread later, memory is often stirred as the expressed emotions of the experience are relived and the cognitive aspects are recalled.

If you plan to collect the journals to determine that students are indeed writing them, make this practice clear from the start. It is impolite (and possibly embarrassing) to tell students that their journals are for their eyes only and then to violate that working principle. (Since in my class journals are graded for length, not content, students are told to place a red mark on any page that they do not wish me to read.)

In-Class Writing

Other important ways that writing can be used in a mathematics education class are more traditional. Most professionals can remember trying to decode notes that seemed clear when taken but were murky when the time came to copy them over. Devoting the last few minutes of class time to composing and comparing notes will avoid this frustrating experience. It will also save time at the start of subsequent classes; help students develop a facility in analyzing, synthesizing, and searching for important relationships; and reduce the amount of time spent in passive listening.

As important as writing is in reviewing what has already occurred, it can be even more useful as a way to focus discussions. One example will suffice. As part of my methods course, students are asked to generate

---

Most of us have had trouble decoding notes that seemed clear when taken.

---

an operational definition of themselves as teachers of mathematics by naming an animal they will be "like." Since metaphorical processing is thought to elicit right, rather than left, brain responses (Ferguson 1980, p. 305), what "comes out" often surprises students, particularly the more verbal or analytic ones. Writing for a few minutes on this "weird" topic and sharing the output helps students understand the many workable styles of teaching. This understanding can, in turn, allow students to develop further their own personal well-thought-out modes rather than look for the "one right way." Asking students to develop such extended comparisons as "How is learning mathematics like learning to ride a

bicycle?" or "What food was your favorite mathematics teacher like?" will stimulate lively discussions and give students a chance to consider a serious topic in an involved but light-hearted way.

It is especially important that teachers tailor explanations to the person(s) being addressed. In writing, this practice is called developing a sense of audience. To help develop this ability in students, you can assign the writing out of a sequence of activities to explain a specific topic, place value for instance, to a second grader meeting it for the first time, to a slow learner in the fifth grade, or to a parent who wants to help a child at home. Individuals or small groups can be asked to create answers for the same question when posed by different people. One possible question is "Why is Jane going to the LD teacher for help in mathematics?" as asked by Jane, her mother, the principal, and the LD teacher. Another question is "Why is it important to learn the number facts?" as posed by a gifted child, a friend who is not in the educational profession, and a member of the school board. Similar questions, when included on tests, are interesting to read. More importantly, they sample more adequately than would an unfocused essay question the student's ability to use knowledge of a specific area flexibly.

Term Papers

Many mathematics educators, believing that prospective teachers must acquire a facility with written language and a familiarity with the literature in the field, require a formal paper. Although positive about the value of this assignment, they look wearily toward the day when they will tote home that set of papers and sigh as they contemplate the hours they will spend marking the inevitable errors in grammar and punctuation. Happily, let me suggest ways to make the task of producing a paper more pleasant for the instructor and more helpful to the student—simultaneously. These methods are (1) *focus on the question, not the topic* and (2) *employ the peer review process*. When I was first ad-

vised that these suggestions would result in dramatic gains in the quality of students' writing, I was skeptical, so I cannot blame you if you share that skepticism.

The first technique is so simple that I felt it could not be very worthwhile. It was simple (and worthwhile) in the way that a safety pin is, and I wondered why I'd never thought of it before. (I took comfort in the fact that the safety pin was not widely known until after 1850.) Once each person in my graduate class had chosen a topic, for example, sex differences in spatial ability, each student was told to identify the question that the paper would answer. Sharing the questions helped individuals decide if they had taken on too broad an area. A revision of the paper's focus was quite easy at this point. In your class, as in mine, someone might not be able to decide what question the paper is to answer. Here is a strong indication that you would have gotten a rambling, irrelevant ten pages or so of interest to no one, not even the writer. You can avoid this unpleasantness by helping such a student formulate a specific and engaging question. Your professional commitment in this process moves from spending a large block of time alone with a large stack of papers—some of which are poor beyond imagining and as difficult to respond to as they were to write—to working individually with students early in the term.

The second part of the formula for better papers is to use the peer review process. Assuming you have established rapport among class members and have helped them (and yourself) understand that criticism can be constructive, the peer review process will prove a valuable addition to your teaching repertoire. During the peer review process, students read each others' early drafts and talk about them. Not only do the authors tend to produce better papers, since more people than "just" you will read them, but in acting as reviewers, students also become familiar with ideas in mathematics education that might not have been covered in class. I use a format for the peer review process developed by the faculty of the Center for Writing at the University of North

Fig. 2
Peer review sheet

Workshop Checklist for _____**'s Paper**

1. Is the introduction clear? _____yes _____no

   Does it create an interest in the topic? _____yes _____no

   Does it reflect both the organizational plan and the central idea of the paper? _____yes _____no

2. What is the thesis? _____

   _____

3. Is the thesis clear and precise? _____yes _____no

4. Is the conclusion clear? _____yes _____no

   Does it restate the thesis of the paper? _____yes _____no

5. How many paragraphs are in the body (excluding the introduction and the conclusion)?_____

   Paraphrase the topic sentence of each paragraph.

   Paragraph A _____

   Paragraph B _____

   Paragraph C _____

   Paragraph D _____

   Paragraph E _____

   Does each topic sentence explain or develop the thesis statement in some way? (Check the ones that do.)

   Are the paragraphs in the best order (well organized)? _____yes _____no

   (Suggest changes.) _____

   _____

   _____

6. Does the author offer evidence to support every major point? _____yes _____no

   If no, which point(s) need additional support?_____

7. Is the style generally graceful, flowing, and pleasant, or stilted, choppy, and difficult to read? _____

   _____

8. Comments:

Name of editor:_____

(Developed by the staff of the Center for Writing, Department of English, University of North Carolina—Wilmington)

Carolina—Wilmington (see figs. 2 and 3). You might want to construct a similar format. (My students knew that I would use this same format in my grading. Thus, I had specific criteria for the As, Bs, and Cs, not to mention the Ds and Fs, I gave.) The process should begin as early as possible in the semester and should include several announced "checkpoints." Since writing is a complex task, as long a time span as possible should be allotted.

Once students have the benefit of one or more peer reviews, it is time for them to consider the comments thoughtfully and revise their papers. It can be helpful for you to draw the distinction between revising (in which words, sentences, or paragraphs are relocated, reshaped, or reworded) and editing (when spelling, punctuation, and other conventions of writing are checked, perhaps against a style manual). Among the poorer papers I have received were those from students who wrote down what they wanted to say and then checked for grammatical errors. Indeed, many thought that good writers produced a final copy on

---

Fig. 3
Rating scale

A. Central Idea

5	4	3	2	1

Very strong. Limited and focused subject. Interesting, original, and perceptive idea, clearly stated. Strong sense of writer's purpose.

Adequate. Interesting idea but either not particularly original or especially perceptive. Clearly stated or implied thesis. Some sense of writer's purpose.

Very weak. Broad, unfocused subject. No apparent thesis, either directly stated or implied. Ideas lack interest, originality, and perception. No sense of writer's purpose.

B. Organization

5	4	3	2	1

Very strong. Interesting, appropriate opening and closing paragraphs. Logical and effective paragraphing. Smooth, clear transitions between and within paragraphs.

Adequate. Appropriate opening and closing paragraphs, although more interest could be generated. Generally effective, logical paragraphing, but some transitions could be smoother.

Very weak. No concern for audience or purpose in opening and closing. Confusing or illogical paragraphing. No transitions.

C. Development

5	4	3	2	1

Very strong. Strong support for thesis. Generous use of concrete details and examples. Well-unified paragraphs.

Adequate. Clear support for thesis, although more development needed for some subpoints. Some details and examples, although a few unsupported generalizations remain. Well-unified paragraphs.

Very weak. No sense of development of single ideas. No use of details or examples to explain generalizations. No unity in paragraphs.

D. Style

5	4	3	2	1

Very strong. Precise, effective word choice. Varied, coherent sentences; minimal wordiness or unnecessary repetition.

Adequate. Generally effective word choice, occasional inexact, trite, or bland phrasing. Some attempt to vary sentences. Occasional wordiness, unnecessary repetition, or unclear phrasing.

Very weak. Word choice lacking in precision, effectiveness, and accuracy. No attempt to vary sentence structure. Excessive wordiness, unnecessary repetition, and unclear phrasing.

E. Technical Control

5	4	3	2	1

Very strong. No distracting punctuation problems. No major errors in grammar (e.g., agreement, pronoun reference, verb endings, etc.). No confusing shifts in tense, person, or point of view.

Adequate. Few distracting punctuation problems or major errors in grammar. No confusing shifts in tense, person, or point of view.

Very weak. Frequent distracting punctuation problems and major errors in grammar. Confusing shifts in tense, person, and point of view. Frequent misspellings. Improper capitalization and abbreviations.

Comments:

---

the first trial. Disabusing inexperienced writers of that notion is an important step in helping them develop a more mature style. Intrepid instructors might even share some early drafts of their published writing. It was a shock for my students to see the amount of crossing out and stapling on my early drafts, I can tell you. (Of course, if you use a word processor, this step is not feasible. Perhaps a look at what word processor programs are designed to do may give students some hints about what authors do in the task of revision.) The value of making these invisible processes visible is suggested by an author who has a lot to say about quality in writing—and in life. "He felt that by exposing classes to his own sentences as he made them, with all the misgivings and hang-ups and erasures, he would give a more honest picture of what writing was like than by spending class time picking nits in completed student work or holding up the completed work of masters for emulation" (Pirsig 1974, p. 187–88).

Many students' writing is boring because only ill-digested bits of information are presented. One gets no feeling that a *person* is involved. If your background in composition is similar to mine, you were forbidden to use "I" or "you" in formal papers. The closer "one" came to a mechanical but well-stated paper, the "better" it was. Now not everyone agrees that quality in writing is synonymous with sterile objectivity (Smith 1982). Many scholars now acknowledge that writing is done by, and for, people about topics of human interest. Authors are permitted, even encouraged, to use "I." Acknowledging personal involvement is the first line of defense against the stultifying separation of the writer from the topic. "At the moment of pure Quality, subject and object are identical" (Pirsig 1974, p. 284). Pirsig goes on to say that "when one isn't dominated by separateness from what he's working on, then one can be said to 'care' about what he's doing. That is what caring really is, a feeling of identification with what one's doing. When one has this feeling, then he also sees the inverse side of caring—Quality itself" (p. 290).

When the peer review process is used, it is nice to follow the lead of published authors and include an acknowledgment page in the final paper (Maimon et al. 1981). On this page the student can thank (and think about) all who have helped improve the paper. Learning to appreciate what others do for us and publicly giving thanks is a nice habit to form.

Ways exist to make the task of producing a paper more pleasant for the instructor and more helpful for the student.

The day comes when the students' drafts have been reviewed, revised, retyped, and edited, and the instructor sits alone to face the twin evils of plagiarism and evaluation (Johnson 1983). The first evil can be vanquished easily if students are required to turn in all early drafts as well as the final paper. You can assure them that neatness does *not* count, except in the final draft. Evaluation will be with us always, but since research reveals that students pay little attention to instructors' comments on returned papers (Maimon et al. 1983), frustration and time can be saved by using a "new" evaluation procedure. With a format similar to that shown in figure 3, you can rapidly and accurately assess the overall quality of work. The points attained can be scaled to arrive at a numeric or alphabetic grade. Frequent grammatical and syntactical errors can be highlighted by copy editing a paragraph or two rather than the entire ten or twenty pages.

A Final Word

"Learning about a subject means more than memorizing axioms, dates and formulas. You need to develop general intellectual skills that will allow you to understand your discipline in its entirety, that is, to approach it intelligently, knowing what questions to ask, where to discover the answers to those questions, and, finally how to develop and organize your own ideas about the subject" (Maimon et al. 1983, p. 18). Surely it is just these procedures that mathematics educators would like to nurture in their students. Including writing in a mathematics education class can help that happen.

References

Ferguson, Marilyn. *The Aquarian Conspiracy: Personal and Social Transformation in the 1980's.* Los Angeles: J. P. Tarcher, 1980.

Johnson, Marvin L. "Writing in Mathematics Classes: A Valuable Tool for Learning." *Mathematics Teacher* 76 (February 1983): 117–19.

Maimon, Elaine, Gerald Belcher, Gail Hearn, Barbara Nodine, and Finbarr O'Connor. *Writing in the Arts and Sciences.* Cambridge, Mass.: Winthrop Publishers, 1981.

Pirsig, Robert M. *Zen and the Art of Motorcycle Maintenance.* New York: Bantam Books, 1974.

Smith, Frank. *Writing and the Writer.* New York: Holt, Rinehart & Winston, 1982.

Teacher Education

How Teacher Educators Can Use Manipulative Materials with Preservice Teachers

By **Sharon L. Young**
Louisiana State University, Baton Rouge, LA 70803

Not long ago I was observing Martha Peterson, a preservice teacher from my methods course, as she was teaching the subtraction algorithm to some third graders. Martha was using base-ten blocks to show how the physical exchange of a "long" for ten "units," in relation to a written step in the algorithm, when Johnny exclaimed, "So that's why you cross out the tens!" As we left the classroom, Martha was elated and said, "Did you hear Johnny? The base-ten blocks really helped him understand. I didn't think manipulative materials would really work, but now I do."

Scenarios similar to this occur frequently when manipulative materials are fully integrated into a methods course. My own experiences have shown me that methods instructors can use a variety of activities to introduce preservice teachers to manipulative materials. Some activities prepare preservice teachers to use the materials and others have the students use the materials in actual instructional situations.

The preparatory activity should focus on the definition of manipulative materials. The following definition is one that I have used with my students: manipulative materials are objects which represent mathematical ideas that can be abstracted through physical involvement with the objects. With such a definition, preservice teachers can evaluate various instructional aids to determine whether or not they are manipulative materials. For example, they might determine that a bead abacus and colored rods are manipulative materials for teaching addition but that flash cards are not.

Many different manipulative materials can be used to teach a single mathematical concept or skill. It is important that preservice teachers learn which materials are appropriate for which of the various levels of mathematical understanding, as some may require children to think in more abstract ways than others. A useful activity is to have students rank a variety of materials according to the level of abstraction required to use them for a particular concept or skill. For example, among place-value materials the easiest to think about are objects that can be grouped into tens by hooking together (plastic connecting cubes) or bundling (wooden sticks). Materials that involve a trading process, such as base-ten blocks, require a greater degree of abstraction in thinking, and colored chips (for chip trading) require a higher degree of abstraction than base-ten blocks.

In the previous activity, the preservice student learned that a variety of materials can be used to teach a single concept or skill. The next activity "reverses" the approach by focusing on the many skills and concepts which can be taught with a single manipulative material. Whenever I introduce preservice teachers to a manipulative material for the first time, I have them use the material with a specific skill or concept, and then I also have them identify other mathematical ideas that can be taught through the use of that material. For example, students can discover that base-ten blocks can be used to teach the four basic algorithms for whole numbers and decimal fractions as well as place value.

It is important that preservice teachers learn how to help children connect the physical manipulation of materials with written symbols. One activity that focuses on building bridges between materials and symbols is to have your students develop specific, step-by-step teaching procedures that will enable a child to see that written symbols are simply a way to record the results of manipulating materials. For example, fraction tiles could be used to show children how to add fractions with unlike denominators on a concrete level and then in written form.

Preservice teachers also need to discover that manipulative materials and textbooks can form a working partnership. The first instructional lesson plan that my students write focuses on this important notion. I have them develop a plan to teach a specific lesson from a current textbook. Manipulative materials must constitute a major portion of the instructional time in the plan, and some link must enable the child (and, therefore, the preservice teacher) to see the connection between the materials and the textbook.

With today's limited school budgets it is important for preservice teachers to learn to develop inexpensive manipulative materials. One activity is to brainstorm ways to make inexpensive models of manufactured materials. For example, paper clips can be hooked together in lieu of plastic connecting cubes, base-ten blocks can be cut from graph paper, and tangrams can be cut from colored posterboard. For a related activity, students· are assigned a specific skill or concept for which they are to design a lesson using items from a ''junk'' box. Another activity is to have students make their own manipulative materials. Beansticks, fraction kits, geoboards, and stick abaci can all be made easily and inexpensively. Students are often amazed at their own ingenuity and learn that inexpensive materials can be as effective as purchased manipulative materials.

Instructors of methods courses are interested in whether or not preservice teachers can effectively use manipulative materials to teach concepts and skills. This concern can be addressed by observing students in peer-teaching situations. Students can prepare lessons that use manipulative materials and teach the lessons to one or more preservice teachers. These microteaching situations can be videotaped so that students can observe their own teaching.

Of all the activities that train preservice teachers to use manipulative materials, the most beneficial one is teaching children. As a part of the field-experience component of the course, I have my students prepare and teach lessons to children in elementary school classrooms. The lessons must incorporate the use of manipulative materials, and they must also fit in with the classroom teacher's ongoing plan for mathematics instruction. Several years ago I did not require my preservice teachers to use manipulative materials with children. When given a choice, many preservice teachers did not use the materials, despite extensive work with them in the methods course. Now that I require the use of manipulative materials in actual teaching situations, most preservice teachers become ''sold'' on their use.

Another useful teaching activity is to have preservice teachers use manipulative materials with children for purposes of evaluation. As an example, the methods students can use materials such as connecting cubes or base-ten blocks to determine how well children understand the concept of addition or the skill of subtracting two-digit numbers. Such evaluation can be diagnostic or for postassessment.

Preservice teachers can synthesize the information and skills they have obtained in the previous teaching activities when they develop a project in which they prepare an instructional unit for a specific skill or concept. The project requires that the student integrate manipulative materials with the textbook or other printed material and use manipulative materials within the framework of long-range objectives. The project can also reinforce the use of inexpensive materials by requiring the incorporation of at least one teacher-made manipulative material.

Teachers must be selective in their choice and use of manipulative materials. Constraints on budgets and classroom space limit the teacher's inventory of such materials. Given such limits, teachers should make their selections based on the age and interest of the children, the appropriate mathematical content, and their own personal teaching styles. A previous article in this journal (Reys 1971) offers additional pedagogical and physical criteria for selecting manipulative materials. In an activity that prepares preservice teachers to select appropriate materials, the methods instructor can have students specify the type and quantity of purchased and teacher-made manipulative materials needed to teach mathematics for a hypothetical class. It is specifically suggested that this activity be used toward the end of the course, as it requires preservice teachers to synthesize and apply their knowledge about manipulative materials.

Using manipulative materials with preservice teachers can be a rewarding experience for the instructor of mathematics methods. The following are some managerial suggestions that will be of help when using materials in a methods course.

• Assign students to work with manipulative materials in groups, when appropriate. Small groups encourage the participation of all students, not just those who are highly vocal. Group decisions can be shared with other groups.

• Seat students around tables when using manipulative materials. This gives needed work space that individual slant-topped desks cannot provide and encourages students to work together.

• Give students time to explore new materials. As with children, students need time to acquaint themselves with materials before they use them for specific purposes.

• Give students access to materials outside of class. This goal can be accomplished through a mathematics laboratory, an instructional materials center, or by having students purchase a materials kit, much as they would purchase a textbook for the course.

• Use one of the many methods textbooks that emphasize the use of manipulative materials.

When teacher educators introduce their preservice teachers to manipulative materials, they should use a comprehensive approach that will maximize the students' opportunities and experiences. The activities described above are a necessarily brief sample of many techniques that can be used to expose preservice teachers to the value of using manipulative materials to teach children.

Reference

Reys, Robert. ''Considerations for Teachers Using Manipulative Materials.'' *Arithmetic Teacher* 18 (December 1971):551–58. ●

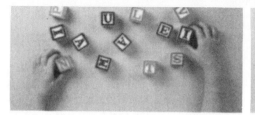
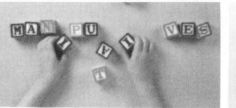

Hands On:
Help for Teachers

By **Cecil R. Trueblood**

The preparation of prospective teachers to use manipulatives in the classroom is a multidimensional task. These dimensions include the following:

1. Selection and use of manipulatives that correctly represent the mathematical concepts they must teach
2. Assessment of children's thought processes as they use manipulatives to form mental images of mathematical concepts
3. Planning and management of mathematics instruction that involves manipulatives

Each of these dimensions will be discussed in some detail within the context of the elementary and early childhood teacher education program at Pennsylvania State University. This context should help the reader understand the nature, scope, and variety of settings and activities required to instruct teachers in the use of manipulative materials with young children.

The students are required to enroll in a mathematics education course, a reading methods course, and a practicum in the same semester, which assures a committed block of time every day for fifteen weeks. The students must also have successfully completed a mathematics course for elementary school teachers. The major emphasis of the mathematics education course is to teach preservice teachers

Cecil Trueblood is involved in teaching and research in mathematics education at Pennsylvania State University, University Park, PA 16801.

how to plan and implement a diagnostic approach to teaching mathematics and how to use a variety of manipulatives in this approach. The instructors plan their activities around common objectives and use a common textbook (Heimer and Trueblood 1978). A well-equipped mathematics education laboratory is available that is equipped with frequently used manipulatives and with tables and chairs rather than the typical desks found in most college classrooms. Available also are video and audio equipment needed to conduct microteaching and

Prospective teachers use manipulatives in their teaching in the same manner in which they are taught.

to record diagnostic interviews. In addition, the instructors have at their disposal an instructional support center that is used to manage via computer the achievement data·on students' performance on tests and competency checks for each manipulative aid presented during the semester.

The public schools in the area have cooperated to the fullest extent possible by providing classrooms, manipulative equipment, and, most important, classroom teachers who help model and support the general goal of the instruction given in the professional coursework. The instructors referred to in this article are all certified, experienced classroom teachers who are doctoral students interested in becoming mathematics educators. They

are supervised by a senior faculty member who is responsible for the quality of instruction and the research and formative evaluation that is a part of the program.

Mathematics Education Laboratory

The mathematics education laboratory (math lab) was developed so that prospective teachers could learn how to use a variety of manipulatives to represent the mathematical concepts, relationships, and operations taught in the elementary school. Another purpose of a math lab is to provide a model teaching environment where uses of manipulatives, computer software, and other curricular materials can be illustrated and to make certain these materials are available for use in practicums and for research purposes. We have found that laboratory instruction that presents the philosophy and child-development theory behind the use of manipulatives, followed by reinforcement in the classroom, has a major influence on how and whether prospective teachers use manipulatives with children. We have also found that prospective teachers use manipulatives in their teaching in the same manner in which they are taught, providing that classroom teachers reinforce and support their efforts.

The commercial and teacher-made manipulatives in the math lab are stored in individual kits so that each prospective teacher can easily locate and use them when required. The kits include such aids as Cuisenaire rods, fraction bars, multibase arithmetic blocks, and teacher-made abacuses. This procedure also helps illustrate what classroom organization and management is required to maximize time on task, one of the problems that must be monitored when children work with manipulatives.

The use of manipulatives is presented within the context of a diagnostic approach to the teaching of mathematics. The instruction in the math lab is designed to illustrate a variety of teaching techniques and strategies that incorporate the use of manipulatives. These strategies focus on how to use manipulatives to teach concepts, relationships, operations, and problem solving. The instruction with manipulatives is conducted individually and in groups of various sizes to illustrate the influence of group size on such instruction.

To provide a credible model of diagnostically oriented instruction, the instructor's first procedure is to administer a series of diagnostic tests to identify what concepts, operations, relationships, and so on, will be taught with the manipulatives in the math lab. Since the tests are generated and scored by computer, each student receives a personal copy of the results of her or his performance, along with appropriate prescriptions. These prescriptions include activities in the math lab, assigned readings, and self-directed activities that help familiarize them with the uses of the manipulatives in the lab.

Because the students must spend a considerable amount of time in the math lab, it is open on weekdays, evenings, and weekends. The math lab is supervised by undergraduate teacher-education students who have work-study grants. These student supervisors are trained by the mathematics education instructors in how to distribute and collect the self-instructional materials and related manipulatives. The math lab and the materials are also available to the prospective teachers to use in schools during their practicums. The availability of manipulatives is critical to whether prospective teachers have the flexibility they need to teach the mathematics content identified by the classroom teacher.

Fig. 1   Matrix of instructional modes

Student output modes for responding to instructional tasks	Teacher input modes for presenting instructional tasks		
	M	G	A
Manipulative (M)	M, M	G, M	A, M
Graphic (G)	M, G	G, G	A, G
Abstract (A)	M, A	G, A	A, A

The general model of instruction used in the math lab during scheduled class sessions consists of three steps. The model is designed so that the students first learn how a particular manipulative is used with respect to familiar mathematics content. Then they experience how that manipulative helps them learn an unfamiliar or forgotten concept, relationship, or operation. The three steps are these:

Step 1: Using a familiar concept, relationship, or operation, the instructor demonstrates how to use a manipulative so that the student becomes familiar with its features and potential.

The use of manipulatives is presented within the context of a diagnostic approach to teaching mathematics.

Step 2: The instructor then provides hands-on manipulative instruction on an unfamiliar concept. The diagnostic-test results mentioned previously are used to identify concepts unfamiliar to the prospective teachers.

Step 3: Following mastery of that concept, the instructor has the students analyze and discuss the rationale underlying the instructional process and what mental imagery the manipulatives helped them formulate.

We have found that this experien-

tial model is the most efficient and effective way to introduce prospective teachers to a manipulative. It also seems to motivate them to do additional work with manipulatives on an independent-study basis.

Following instruction on how to use manipulatives as instructional cues, the students are introduced to a matrix of instructional modes that applies Bruner's (1964) modes of representation. Figure 1 presents this application. The cells in the matrix are used to classify instructional tasks according to the mode used by the teacher in presenting a task to students (teacher input) and the mode used by the student to respond to that task (student output). The matrix gives the prospective teachers a more precise definition of nine possible instructional modes, five of which involve the use of manipulative materials by the teacher or the student. Converting the cells of the matrix into an observational checklist is also a useful way to balance instruction in manipulative and symbolic or abstract activities.

Figure 2 is an example of a completed checklist matrix that indicates modes of representation used by a teacher. The cells of the matrix contain the observer's notes that describe the mathematics content, the examples provided by the teacher as input, and the ways in which the teacher asked the students to respond. Following observations with such a checklist, prospective teachers usually conclude that without careful planning, only symbolic and graphic modes of instruction tend to be used.

The matrix is also a useful tool for helping students plan for the use of

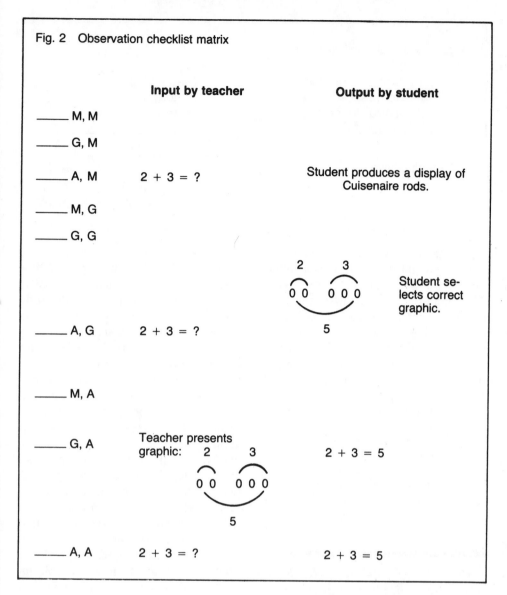

Fig. 2 Observation checklist matrix

manipulatives in their practicums. It also assists them in sequencing their lessons from one day to the next. It is important to point out that our mathematics education instructors are also the practicum supervisors. This staffing procedure was initiated to help insure that the instruction given in the math lab is carried over into the practicum in a manner acceptable to the cooperating teacher. The staffing procedure also gives the supervisor and prospective teacher common points of reference and a common vocabulary with which they can critique instruction with children. This procedure also helps reduce the cost of staffing.

How Children Think about Mathematics

Most prospective teachers have little

or no firsthand experience in observing or listening to a child explain his or her understanding of a mathematical concept or relationship. In the math lab, prospective teachers are also

---

It is important to learn what to ask and what not to ask in a diagnostic setting.

---

shown methods of conducting and analyzing the results from interviews. Here they are taught how to conduct a diagnostic interview with manipulatives to tap a child's perceptions and mental imagery.

This instruction involves students' exposure to the manipulatives previously discussed, as well to observa-

tions and discussions of videotaped diagnostic interviews. Preparing and critiquing an interview administered to peers has also produced useful results. The important competencies taught using these two techniques are knowing what type of questions to ask (e.g., checking for understanding) and *not* to ask (e.g., cues that hint at possible correct answers) and selecting and presenting manipulatives in a diagnostic rather than instructional manner during the interview. Another important outcome is learning how to analyze the results of the interview so that valid interpretations of children's responses are made. It is also important for students to be able to use the results of the interview to describe what mental images children are using to represent a particular concept, relationship, or operation.

The Practicum

The intensive practicum begins at the end of the tenth week of the semester, after the instruction in the mathematics lab is completed. Students work in a classroom five half-days a week for five weeks. The fifth day, Friday, is used for reflection on the week's learning experiences and instruction. During the practicum, students serve as a teacher's aide and teach mathematics to individuals or small groups under the guidance and clinical supervision of their university instructor and a classroom teacher. More specifically, the practicum consists of planning, teaching, and evaluating mathematics lessons using the diagnostic approach presented in the math lab.

On the basis of guidance from the cooperating teacher, the prospective teacher selects or designs instruction for small groups around objectives that include manipulative, graphic, and symbolic activities. The evaluation of the student is based on the supervisor's and teacher's observations. This conference, which includes the prospective teacher, the cooperating teacher, and the supervisor, focuses on assessment of the appropriateness of the instructional activities used and whether the objectives specified for these activities were met.

Research and Formative Evaluation

To identify which instructional activities are most effective in preparing prospective teachers to use manipulatives, ongoing research and formative evaluations are required as integral parts of a teacher-education program. To support and encourage the participation of classroom teachers in such research and evaluation, it is important to disseminate the results to the participants and to the profession at large. We have found that without such feedback, professionals who must make a heavy investment of time and energy are likely not to be supportive of the practicum activities described in this article.

Some research studies have been conducted at Penn State on the use of manipulatives in teacher-education practicums and with children. These studies are listed at the end of this article. The results indicate that prospective teachers' attitudes toward mathematics and the use of manipulatives in mathematics instruction should continue to be a concern to those interested in research and in increasing the use of manipulatives with children. It also appears that weighting instruction in the primary grades toward manipulative activities should be more of a concern than in the upper intermediate grades, where a balance of manipulative, graphic, and symbolic activities seems more advisable.

Prospective teachers resist using manipulatives in the classroom for two reasons: a lack of confidence in their own ability to use manipulative materials correctly and the general belief that children will become too dependent on these materials and, as a result, will not master basic computational algorithms and related concepts. This general belief seems related to a lack of confidence in helping children make the transition from the concrete to the abstract. We have observed that participating in the practicum experiences has increased both prospective and cooperating teachers' use of manipulatives with children.

We have also found that the actual use of manipulatives with an individual child in a tutorial situation seems to produce the most change in prospective teachers' negative attitudes toward manipulative materials. This change in attitude seems to come from two sources. First, using manipulatives with a single child in a tutorial situation highlights the influence teachers have on helping a child learn mathematical concepts and relationships. This influence is communicated by the child's increased ability to describe the structure of a concept and his or her increased attention span and motivation to learn. Second, using manipulatives with individual children or small groups provides more immediate and concrete feedback

---

Using manipulatives with an individual child improves prospective teachers' attitudes toward manipulatives.

---

than does working with symbols in a lecture-demonstration situation.

Another important influence on prospective teachers' attitudes is the amount of practice they have in using manipulatives and in the self-directed activities used to demonstrate their competence with each manipulative. This influence is also related to their perceptions of how manipulatives help them improve their understanding of mathematical concepts and relationships taught in the elementary school.

Our formative-evaluation results show that the diagnostic approach used in this course helps improve prospective teachers' understanding of, and attitude toward, mathematics. Diagnostic teaching experiences have also helped reduce the anxiety associated with using manipulatives with children. Finally, we believe our experience helps confirm the role that math labs and practicums should have in preparing teachers of elementary school mathematics to use manipulatives.

For those interested in developing teacher-education experiences similar to ours, one way to start is to adopt or adapt the general three-step model of instruction used to introduce manipulatives in the math lab. This model would fit almost any type of program configuration and hence could serve as a point of departure for teacher educators who want to prepare their prospective teachers to use manipulative materials more effectively.

References

Bruner, Jerome S. "Some Theorems on Instruction Illustrated with Reference to Mathematics." In *Theories of Learning Instruction*, Sixty-third Yearbook of the National Society for the Study of Education, edited by E. R. Hilgard. Chicago: University of Chicago Press, 1964.

Heimer, Ralph T., and Cecil R. Trueblood. *Strategies for Teaching Children Mathematics*. Reading, Mass.: Addison-Wesley Publishing Co., 1978.

Research Studies

Attivo, Barbara A. "The Effects of Their Instructional Strategies in Prospective Teachers' Ability to Estimate Length and Area in the Metric System." Ph.D diss., Pennsylvania State University, 1979.

Fennell, Francis, and Cecil R. Trueblood. "The Elementary School as a Training Laboratory and Its Effect on Low Achieving Sixth Graders." *Journal for Research in Mathematics Education* 8 (March 1977):97–106.

Fernsler, Thomas. "The Evaluation of Two Types of Instructional Strategies on Preservice Elementary Teachers' Attitudes toward Hand Calculators." Ph.D diss., Pennsylvania State University, 1983.

Gonzales, Angela. "Effectiveness of Problem-solving Activities in Changing Perceived Elementary Teachers' Attitude toward Mathematical Problem Solving." Ph.D. diss., Pennsylvania State University, 1983.

Houser, Larry L., and Cecil R. Trueblood. "Transfer of Learning on Similar Metric Learning Tasks." *Journal of Educational Research* 68 (February 1975):235–37.

Kongsasana, Prasit. "Metric Attitude and Achievement of Preservice Elementary Teachers as a Result of Three Instructional Approaches." Ph.D. diss., Pennsylvania State University, 1978.

Myers, James R. "Change of Attitudes toward Mathematics of Prospective Elementary Teachers in a Mathematics Methods Course." Ph.D. diss., West Virginia University, 1983.

Smith, Susan R., Michael Szabo, and Cecil R. Trueblood. "Modes of Instruction for Teaching Linear Measurement Skills." *Journal of Educational Research* 73 (January/February 1980):151–54.

The author wishes to acknowledge the important role played by the teachers and administrators from the Penns Valley, Bellefonte, and State College area school districts in the preparation of teachers at Pennsylvania State University.

A mathematics laboratory for prospective teachers

D A V I D   M .   C L A R K S O N
Syracuse University, Syracuse, New York

*David Clarkson recounts one of his projects as a
visiting lecturer at the State University College at New Paltz,
New York. At the present, he is associate director of the
Madison Project and is teaching at Syracuse University.*

The Report of the Cambridge Conference on the Correlation of Science and Mathematics in the Schools recommends that schools of education plan programs of "apprentice teaching in the schools, including work with materials of the sort being developed in new curriculum projects."[1] A group of mathematics educators in England has urged the use of courses emphasizing problem solving: "It is the exploration of these more open problems which we feel to be the essential characteristic of real mathematical activity."[2] A loud chorus of opinion suggests that courses in methodology should be jointly planned and executed by both mathematicians and educators and that they should involve practical work with children. When the opportunity to design an experimental elementary mathematics methods course was offered the writer, he decided to emphasize the mathematics laboratory approach which gives an important role to

problem solving. Conferences with members of the mathematics and education departments, as well as with school officials, paved the way for the experiment; the sympathetic support of the chairman of the division of education at the college made it possible financially.

A few teachers in neighboring school districts had been using activity methods in their classes. Because there were not enough of them to provide an in-school laboratory for the course sections, it was decided to bring children onto the campus. A block of time—most of a morning—was reserved in the college schedule so that every college student would have a substantial experience working with a child. The children came from four schools, including a residential school for delinquents. Their ages ranged from five to twelve years. Each child, and each college student, had an opportunity to participate in from five to ten laboratory sessions during the semester. No attempt was made to create a typical elementary classroom scene; rather, about sixty children and an equal number of students worked together in three or four college classrooms, halls, and outdoors. The sessions constituted about one-third of the work of the college course.

---

[1] Cambridge Conference, *Goals for the Correlation of Elementary Science and Mathematics* (Boston: Houghton Mifflin Co. for Education Development Center, 1969).

[2] Mathematics Section of the Association of Teachers in Colleges and Departments of Education, *Teaching Mathematics, Main Courses in Colleges of Education* (London: A.T.C.D.E., 1967).

Theoretical discussions in the college course were always closely followed by practical work with materials in class, and subsequently with children. Because of time limitations many topics could not be included. Students were encouraged to develop their own interests and to pursue topics in depth rather than to survey the field superficially. Behind this decision was the belief that if a student has really worked out the problem of discussing measurement with a child, say, he may develop skills and approaches that will enable him to do a similar job with many other topics of elementary mathematics. If a good method of attack is gained by several such experiences it should be applicable to many more. In any event, there simply isn't enough time in a semester course to cover even a majority of the current topics.

The mathematics laboratory approaches that formed the focus of the course have only recently come into vogue in the United States. While they have some roots in the developmental psychology of Piaget and others, they have also developed in response to the heuristic, as opposed to the formalist, school of mathematics educators. One of the best statements of this viewpoint may be found in the book *Freedom to Learn*.[3] A part of the early meetings of the course was devoted to a discussion of Piagetian tests, particularly those related to the idea of conservation. The college students then had opportunities to administer some of these tests to children in the laboratory setting and to look critically at them. A few became so fascinated with the results that they devoted part of their vacation time to testing children in their neighborhoods, and several worked up major reports on this aspect of the course. A video-tape recorder was used during laboratory periods to record some of this activity so that it could be shared with the rest of the students later.

Some mathematics educators are worried that an excessive emphasis on materials—messing about with "things"—will detract from the development of mathematical content in the laboratory situation. Some of the activities, such as playing with "Tangrams," may seem to bear only a trivial relation to the study of serious mathematics. A major effort of the course was to relate the activities to significant mathematics, but this was done in the informal context of the laboratory. A variety of texts was used; the Nuffield Guides and other recent publications were particularly helpful.[4, 5] Students in the course kept logs of each session at which they worked with children, and these were read and commented upon by the teacher. For purposes of evaluation, the logs proved to be even more valuable than the usual "projects." Motivation of both students and children was extremely high and was maintained throughout the course. The following selection of excerpts from student logs conveys some of its flavor:

One girl and I took a yardstick and we measured anything that she wanted in the building. She liked to measure long things because then she could add the numbers. It was interesting to note that she liked to add the numbers but she was too busy to write them out on a chart. She enjoyed the chance to move about the building freely but disliked the idea of sitting down and working with the information she had acquired. But then this little girl noticed that another child was putting a chart up for display and she wanted me to make one for her. We compromised, and I wrote the words while she completed the mathematical details. When she was done she really seemed to get enjoyment because she could see the comparison between the [measure of the] door size and the water fountain.

In the lab I noticed a child's response to the free atmosphere. Three boys were working with tangrams. Two of the boys were very persevering in their attempt to form a square and worked at it for more than half an hour even though they could easily have chosen something else. The third boy just could not catch on to the tangrams and without any embarrassment or feeling of failure he was able to go to another activity.

[3] Edith E. Biggs and James R. MacLean, *Freedom to Learn: An Active Learning Approach to Mathematics* (Reading, Mass.: Addison-Wesley, 1969).

[4] Nuffield Mathematics Project, *Guides* (New York: John Wiley & Sons, 1968).

[5] Association of Teachers of Mathematics, *Notes on Mathematics in Primary Schools* (New York: Cambridge University Press, 1967).

Some bright eighth grade boys wanted to make graphs and, since I was free, I agreed to help them although I don't know much about graphing. We went to a building on the campus which has an elevator and used a stop watch to collect data on its motion. The boys thought of collecting data that I would have ignored. For example, they compared the time it took to get from floors 2 to 3 with the time it took to go from 3 to 2. We were all surprised to find it takes longer to go down than it does to go up. Afterwards they made graphs of the data and were perplexed to find that only insignificant differences showed up. The next session they experimented with changing the scales on the graphs and found that this could make differences *appear* significant. By letting them take the lead I learned a lot!

The content of the course and the laboratory sessions accompanying it was eclectic and open-ended. Structural materials such as Cuisenaire rods, multibase blocks, attribute blocks, and student-made materials were available, as were a number of suggestions for activity cards. Much work was done with nailboards and shapes. Graphing activity was everywhere in evidence. Balances, tape measures, stopwatches, and other measuring instruments, including a spate of homemade trundle wheels, got extensive play. Some of this equipment was expensive, but many cheaper substitutes could have been made had there been more planning time.[6] The total cost of a well-equipped mathematics laboratory at the elementary level should be less than $500, and can be considerably less. Furthermore, most of the expensive items are permanent acquisitions. Students were encouraged to make their own materials and, where appropriate, to involve the children in this also.

Five basic content areas were developed: graphing, measurement, geometrical relations, number patterns, and reasoning. Of course, these areas overlapped. For example, in the work with graphs, students and children progressed from simple charts based on counting (histograms, etc.) to empirically derived graphs (spring stretch, ball bounce, etc.) to graphs of functions (guess my rule, etc.). Questions of interpolation and extrapolation were raised, as well as simple concepts of analytic geometry. Functions relating the number of nails on a nailboard to the areas of shapes stretched on it (Pick's theorem) and the functions that emerge from games were also discussed. Opportunities to strengthen computational skills and explore the structures behind those techniques were not ignored, but no attempt was made to develop conventional lessons in the skills.

There are obvious defects in a program of teacher education which waits until the students are in their last two years of college to give them direct exposure to work with children. Moreover, many educators now question the value of student teaching when it is not preceded by extensive experience with individual pupils and small groups. For most prospective teachers, the job of classroom organization and management is all-consuming. Little effort may be reserved for the kind of observation, analysis, conversation, and evaluation which comes naturally when one student is working with one child. We speak of the values of sensitivity training, and particularly of the close observations of individuals this implies, but we often fail to make enough provision for this kind of experience. Even further, we speak of instructional objectives often without giving our college students sufficient opportunity to try them out in the microcosm of a one-to-one confrontation. It was to meet these obvious needs that the laboratory sessions were organized on a one-to-one basis.

Because of the "free" atmosphere of the sessions, many students had a chance to work with children of widely varying ages and abilities. Some of them discovered they preferred to work with younger or older children before they were locked into a student teaching assignment that might have proved uncongenial for them. Some students had a chance to match their abstract idealistic desire to teach in the big city

6 Patricia S. Davidson, "An Annotated Bibliography of Suggested Manipulative Devices," THE ARITHMETIC TEACHER 15 (October 1968): 509–24.

ghetto with their experience of trying to communicate with just one child temporarily removed from that environment. (Happily, in most cases, the experience increased their desire to serve education in this capacity.) It was mainly because of these advantages that the alternatives of using children of one age level, or background, or ability, or by classes were rejected for the program. Had transportation not been available for the children, the plan would have been to send students into schools to work with individual children in the "back of the room" if possible, or in some other informal situation. In our local case, it would have meant placing students in classrooms where activity methods were already being used, although there is still a shortage of such places.

An informal evaluation of the experimental program proceeded on several levels. Most obviously it was evaluated, and positively, by the children and their parents. There was never any difficulty in obtaining the children; they were always "waiting for the bus" on lab mornings. Participating administrators fed back favorable reports from parents and teachers, and some of the latter became interested in the laboratory activities themselves. Since the children were involved for only a few sessions, it was not feasible to attempt any substantial evaluation of what they learned. Anecdotal records and observations of the instructor indicated some increase in attention span, particularly among the academically deprived children. The general consensus of the administrators, teachers, students, college colleagues, and occasional visitors to the lab sessions was that it was certainly not a harmful experience for the children who missed perhaps a half-dozen of their regular morning programs in order to attend. The college students were demonstrably grateful for the opportunity to do some real teaching before they faced the moment of truth with their first class in student teaching. Their observations and techniques improved during the course, but, perhaps most important, they began

to work harder at mathematics as they discovered the need while attempting to keep up with their eager pupils.

This brief account is by no means intended to convey the impression that we didn't make mistakes; far from it! For example, in the first semester trial, the students and children were put together very early in the course before the students had had enough time to become familiar with the materials. This was not a disadvantage to students who were strong in their self-concept as teacher, but it was traumatic to some students who found themselves behind the children they were working with. Occasionally the classes were rather chaotic, and some children wasted quite a bit of time before they caught on to the lack of externally imposed discipline. Record keeping left something to be desired at the beginning, and the instructor was so busy with the over-all scene that some students were denied direct evaluation during the laboratory periods. The administrative detail, particularly getting the pupils on and off campus, was extensive at first. Advance planning was necessary to obtain a workable schedule, and it took perseverance to keep it. Yet, for all these faults, the program was sufficiently successful to warrant continuation and the recommendation for expansion and adaptation to other circumstances and subject areas.

Much more remains to be learned about both the use of laboratory methods in elementary school mathematics instruction and in teacher preparation programs. One thing is clear, however, from even this small venture: It can work. A half year after the first laboratory sessions one of the students in the course called the instructor long distance to tell him that she was engaged in student teaching and was introducing laboratory methods in her class. The reception of her effort was so positive that she had been asked to help the school's curriculum committee prepare an order for laboratory equipment so that the other teachers could introduce the laboratory method the following year.

Preservice laboratory experiences for mathematics methods courses

CHARLES H. D'AUGUSTINE

A member of the faculty of the College of Education at Ohio University in Athens, Ohio, Charles D'Augustine teaches field-based courses in mathematics education. Students in these courses have part of their methods-course experiences in rural schools in Appalachia.

A criticism made frequently by students in colleges of education is that their methods classes are not sufficiently related to work with children. They have also expressed a desire for work with children earlier in their professional training. The program described in this article was an attempt to answer these criticisms and to provide laboratory or field experiences not normally available in mathematics methods courses. The described activities were developed with junior-year college students who were from one quarter to one year away from senior-year student teaching.

During the 1971–72 academic year, the author introduced the following types of experiences in his preservice elementary methods sections:

1. Sensitivity field trips
2. Behavior sensitization activities
3. Field trips emphasizing the role of decision, values, and skills in the problem-solving processes
4. Peer teaching with methods demonstrated by the instructor
5. Laboratory experiences in which the students teach to children concepts introduced by the instructor.

Sensitivity field trips

In an effort to expand the students' perceptions, to make them more aware of the potential for developing mathematical concepts with familiar objects, and to develop an appreciation for applications of mathematics, the classes were taken on *sensitivity field trips*. These trips consisted of walks along predetermined routes, with periodic stops to discuss the mathematical concepts that could be developed with things seen along the way. The students then were asked to prepare questions or problems that could be used with a class of children to develop mathematical concepts. Some of the problems identified on these walks are listed in table 1.

Table 1

Source of problem	Problems
A plot of grass	How could the number of blades of grass be estimated? (Discussion centered on how samples could be used to make this determination.) How could you determine how much oxygen is given off by the grass?
A manhole cover	Why is a manhole cover circular in shape? Why would a square shape not be a good design for a manhole cover?
A chain link fence	How many aluminum slats of a given length would have to be purchased in order to screen out a section of chain link fence between two poles? (Discussion included the measurement skills the child would need in order to be able to solve this problem, and the teacher's role in helping a child become a problem solver.)

The sensitivity field trips were not ranked highly by the students in comparison to the other types of activities described in this paper. An analysis of the students' reasons for the low rankings did not reveal any common complaint, but the following were among the negative comments reported:

1. Too much class time was spent compared to the value derived from the activity.

2. It would have been better to give several examples in class before beginning the field trip, rather than attempting to start from scratch each time.

During the 1972–73 academic year, this aspect of the program was modified to include not only group sensitization activities but individual field trips. In the latter, individual students go out to find some object or thing that can be used to develop mathematical concepts. Their experiences are then shared with their classmates.

Behavior sensitization activities

The author believes that there are three major components involved in being a successful teacher. Two of these three components, which have long been advocated as part of the training of prospective teachers of mathematics, are knowledge of content and the ability to apply appropriate methods. The third component, while recognized as a desirable quality, has largely been ignored in teacher training. If a teacher possessed the quality, it was usually not because his college training had modified his behavior to this end; rather, it was an accidental product of his experience. This third component has been given many names—a psychologically mature person, a person who can relate to others, or a person skilled in human relations.

In an attempt to get students to discover and consider what behavior they and their classmates might exhibit under stress and when confronted with typical classroom management problems, a series of classroom simulations was planned. It was hoped that by having each student observe and then evaluate his own behavior under stress, some progress in modifying, or at least in suppressing, undesirable behavior could be made.

The following are representative of the simulations used:

1. A teacher is confronted by a student who constantly finishes his work long before the rest of the class.

(One undesirable behavior exhibited was the tendency for the student playing the role of the teacher to assign more of the same type of work or practice to this child.)

2. A teacher is confronted by a disruptive student.

(One undesirable behavior exhibited was the tendency to use practice work as a control technique.)

In the students' evaluation of this aspect of their experience they expressed a desire to participate in an expanded program; they felt that the thrust of this activity would help them in their teaching.

During the 1972–73 academic year, this type of activity was taken over by trained counselors under a program called STOP (Sensitivity to Ourselves and People).

Problem-solving field trips

Too many elementary school teachers view problem solving as being those "story problems" found in a basal series. Although the author recognizes that these story problems play a distinct role in acquainting a child with situations requiring certain skills, story problems are utterly devoid of some basic elements of real problem solving. The delimiting process, the role of values, and the nature of decision-making are some of the elements most commonly omitted by story problems.

In order to make students more aware of the nature of the problem solving processes and of how they might give their own students the experiences necessary to foster growth in problem-solving skills, *decision-making field trips* were instituted.

For example, each student was given three nickels and instructed to go into town to purchase an item for a three-to-five year old child. The student was given a sheet on which he had to record certain information, answers to questions like the following:

a. How has the problem been delimited for you?

b. How have you delimited the problem?

c. What need did you identify that influenced the type of item you decided to purchase?

d. Why did you select the store you did?

e. How did you know where to go in the store?

f. What skills did you use in making the purchase of the items?

g. How would your decision-making processes have been modified if you had been given one dollar? A thousand dollars?

h. What values influenced the selection of your item?

In addition to the decision-making type of field trip, students participated in a measurement-oriented field trip which consisted of the creation of a unit of measure; the establishment of a scale; the making of direct and indirect measurements; the conversion from the nonstandard unit to a specified number of standard units; and the development of a type of measure based on some relationship between existing and specified types of measures.

Student evaluations of these problem-solving field trips were favorable. In reflecting back on this type of experience, the author feels that the problem-solving experiences need to be expanded to include activities such as the translation of problem situations to simplified models, gathering and interpreting data, and construction of a mathematical model from gathered data.

Peer teaching

Although not tested statistically, the effect of peer teaching was most notable in the improvement in performance on meeting behavioral objectives for the course. Peer teaching was limited to reteaching concepts which had been taught by the instructor. Two of the peer teaching experiences are listed in table 2.

Table 2

Concepts	Methods and modes
Tens $\times$ tens	Each student taught one other student tens times tens using a step-by-step analysis which focused on the idea that tens times tens results in hundreds. For example, $$30 \times 60 = (3 \times 10) \times (6 \times 10)$$ $$= (3 \times 6) \times (10 \times 10)$$ $$= 18 \times 100$$ $$= 1800$$
$a \div b = c, b \neq 0$	Each student taught one other student. Using a partition approach and a set of objects, a student taught a sequence of activities that involved going from a divisor greater than one to a divisor of one. For example, $6 \div 3, 6 \div 2, 6 \div 1$.

Mastery of behavioral objectives in terms of time and proficiency was markedly improved when compared with previous classes who had attempted to master the same behavioral objectives without the benefit of peer teaching. (No attempt was made, other than through student evaluations, to evaluate the effectiveness of this type of activity. Increased competency was apparent, but further controlled research is needed to evaluate this part of the pilot study.)

Laboratory experiences with children

In order to have the preservice students get experiences with children which were correlated to their mathematics methods instruction, the following types of activities were initiated:

1. During the course of the quarter, specific techniques were identified and stressed.

Involvement: Sometimes referred to in the literature as "every pupil response technique." Example: Three tens and four ones are placed on the board. The class is asked to hold up a numeral which tells how many ones are on the board. (Oral and physical involvement activities are also included in this category.)

Analogies: The technique of using a story to parallel the development of a concept. Example: Once there was an alligator which would eat the numeral that named the greatest number. Which numeral would he eat, 5 or 7? Correct, he would eat the 7. We show his mouth like this $5 < 7$. We read this "five is less than seven." This analogy is being used to introduce the *less than* and *greater than* symbolism. Students are taught the advantages and disadvantages of using this technique in teaching.

Models: Students are taught some of the advantages and disadvantages of using models as well as some cautions to observe in their use. Example: Number line model.

Discovery: Two types of discovery experiences are developed, discoveries resulting from a sequence of activities and discoveries resulting from pairs of activities. Example: Paired activities suggesting the commutative property of multiplication. What multiplication is suggested by this array?

What multiplication is suggested by this array?

Did we change the number of dots when we turned the array? Is $2 \times 3 = 3 \times 2$?

Translation: Involves using previously learned information to learn new information. Example: Using the fact that a child knows two plus three equals five to learn two tens plus three tens, and two hundreds plus three hundreds.

Modified experiments: Involves the class speculating on an answer, gathering data to determine correctness of suppositions, and modifying the speculations. Example: Children speculate on about how many centimeters it would take to be equal in measure to one yard. Data is gathered.

Definitions: Students are taught the role of definitions in learning to discriminate between things.

Rules: Students are taught the summarizing role of rules in developing concepts.

Analysis: Students are taught how to give step-by-step presentations that emphasize meaning.

2. Prior to their working with children, the class was taught mathematical concepts in such a way as to illustrate the techniques which could be used with the children. (For example, long division using models, involvement, analysis, and analogies.)

3. Each student was instructed in the techniques of giving enrichment, remediation, and follow-up to the teaching of concepts.

4. Each student was assigned one child to observe while their instructor taught an introductory lesson to a group of children for a period of approximately fifteen minutes.

5. Each student began working with his assigned child as soon as the brief introductory lesson was completed. The student was responsible for diagnosing whether the child needed remedial instruction, follow-up on the introductory lesson, or a related enrichment activity, and then for proceeding with the appropriate activity. These tutorial activities were observed by the instructor.

6. After fifteen to twenty minutes of tutorial-type activities, the students returned to their campus classroom and critiqued their successes and failures.

The author feels that there are several distinct advantages of this type of group experience over the standard approach of assigning one student to one classroom.

1. A wide variety of teaching modes can be demonstrated.

2. There is less likelihood of the students being exposed to unidentified poor models of teaching than in a one-classroom-one-student type of laboratory assignment.

3. It is more economical in time and effort for the instructor to have a whole classroom of college students working at one location.

Summary

In attempting to solve problems in the area of mathematics instruction, the author has been testing a wide variety of laboratory experiences. Some have been designed to develop in students a feeling for the nature of mathematics, others to make students aware of the skills they need to develop in teaching mathematics, others to give the students opportunities to practice their teaching skills, and others to help students learn to relate to children and fellow workers at a rational level.

Teacher Education

Mathematics Methods in a Laboratory Setting

By **Wade H. Sherard III**
Furman University, Greenville, SC 29613

In our sequence of mathematics courses for the elementary school teacher, mathematics content is integrated with discussion of teaching methods and materials. We have found that the methods portion of these courses, which concern the teaching of informal geometry, measurement, probability, and statistics, can be taught very effectively in a math lab setting, where the emphasis is on active involvement of the student and on the use of concrete, manipulative materials.

At the beginning of each term we divide our classes of preservice teachers into teams of two, three, or four students. Each team is assigned a rather broad mathematical topic similar to those in the following list:

1. Teaching linear measure, with emphasis on the metric system

2. Teaching liquid measure (capacity), weight (mass), and volume, with emphasis on the metric system

3. Teaching geometric concepts through paper folding

4. Teaching geometric concepts with geoboards

5. Teaching geometric concepts with pattern blocks

6. Teaching geometric concepts with tangrams

7. Teaching geometric concepts with mirrors and the Mira

8. Teaching geometric concepts through transformations (slides, flips, and turns)

9. Teaching geometric concepts through tessellations

10. Teaching the meaning of area and volume

11. Teaching introductory concepts from probability

12. Teaching introductory concepts from statistics

Each team is to research its topic carefully and is challenged to discover the many ways that its assigned topic can be taught to children or the many different mathematical topics that can be taught using the assigned teaching aid or material.

In researching its topic, a team is expected to use traditional, commercial textbook series, innovative text materials (e.g., CEMREL's Comprehensive School Mathematics Program), appropriate manipulative materials, and methods books that are housed in our mathematics lab. Teams are also to collect ideas from available resources in the library. (The *Arithmetic Teacher* and other NCTM publications, for example, are used extensively as resource materials.) Although the topics are assigned at the beginning of the term, most teams usually wait until about two weeks before their topics are scheduled for presentation to complete their reading and research.

When a team has finished its reading and research, it designs and makes a sample of activities or experiments that can be used to teach mathematical concepts from the assigned topic to elementary school children. This sample should include activities for development as well as for reinforcement and drill. For a specific teaching aid or material, the sample should include different activities that illustrate the variety of concepts that can be taught with that aid or material.

The team must then organize, sequence, arrange, and set up its collection of activities in our math lab so that all the other students in the class can have the experience of doing each activity or experiment. Math lab sessions, during which a class actually does these activities, always generate excitement and enthusiasm. The students who have organized and set up the lab get the experience of being teachers, and the other students in the class get the experience of learning and doing mathematics in an activity-oriented setting. Most students who work through the activities do so without any difficulty. The team members who have designed the activities, however, are available at their respective stations to give instructions and to answer any questions that may arise. They often provide copies of their worksheets and materials for the other students in the class to keep for their own files.

To ensure that a team has adequately researched, organized, and planned its topic, we require that it consult with the instructor several days before it is scheduled to set up activities for a math lab session. The team discusses its research efforts with the instructor and describes the activities that will be presented and how they will be organized in the lab. At that time we can point out any critical omissions in the research, correct any errors or misunderstandings, and make suggestions concerning any organizational problems.

This approach to the teaching of methods in a mathematics course has proved to be highly effective for the following reasons:

1. The math lab approach reinforces the learning of mathematics content in areas that preservice teachers find difficult (especially geometry and probability). Students often comment on how much better they understand mathematical concepts after they have worked through several lab activities involving that concept.

2. Preservice teachers are given opportunities to develop and use good problem-solving techniques, such as collecting data, making observations, searching for patterns, and formulating generalizations.

3. The math lab approach emphasizes active learning and the use of concrete, manipulative materials. Teachers are much more likely to use this method of teaching in their own classrooms if they have had similar learning experiences as preservice teachers.

4. Preservice teachers get the opportunity to explore the many ways that a specific topic can be taught as well as the many concepts that can be taught with a specific teaching aid.

5. By doing laboratory activities or experiments, preservice teachers get to experience mathematics integrated with, or applied to, topics from other academic disciplines, such as the sciences or the social sciences.

6. Preservice teachers learn to supplement traditional textbook material by researching and using other sources of ideas. They get useful experience in identifying teaching objectives, selecting or designing activities to teach those objectives, making teaching materials, writing clear, readable instructions, working with others in a small-group setting, and organizing a classroom to use its physical space most effectively.

7. Preservice teachers have a chance to use their imaginations and to be creative in a teaching situation.

I believe that this approach to the learning of teaching methods generates considerable interest and enthusiasm while it provides valuable teaching experiences that preservice teachers need before they enter the classroom.

Bibliography

Bitter, Gary G., Jerald L. Mikesell, and Kathryn Maurdeff. *Activities Handbook for Teaching the Metric System*. Boston: Allyn & Bacon, 1976.

CEMREL. Comprehensive School Mathematics Program, Experimental Version. St. Louis, Mo.: CEMREL, 1976–80.

Foster, T. E. *Tangram Patterns*. Palo Alto, Calif.: Creative Publications, 1977.

MINNEMAST. Coordinated Mathematics-Science Series. Minneapolis: Minnesota Mathematics and Science Teaching Project, 1971.

Olson, Alton T. *Mathematics through Paper Folding*. Reston, Va.: National Council of Teachers of Mathematics, 1975.

Pasternack, Marian, and Linda Silvey. *Pattern Blocks: Activities A* and *Activities B*. Palo Alto, Calif.: Creative Publications, 1975.

Phillips, Jo McKeeby, and Russell E. Zwoyer. *Motion Geometry, Book 1: Slides, Flips, and Turns*. University of Illinois Committee on School Mathematics. New York: Harper & Row, Publishers, 1969.

Ranucci, E. R., and J. L. Teeters. *Creating Escher-Type Drawings*. Palo Alto, Calif.: Creative Publications, 1977.

Seymour, Dale. *Tangramath*. Palo Alto, Calif.: Creative Publications, 1971.

Shulte, Albert P., and James R. Smart, eds. *Teaching Statistics and Probability*, 1981 Yearbook of the National Council of Teachers of Mathematics. Reston, Va.: The Council, 1981.

Resource Materials and Equipment

Assorted metric measuring devices, commercial and homemade

ETA Beginners' Metric Kit from Educational Teaching Aids

Geoboard Activity Cards, Primary Set and Intermediate Set, from Scott Resources

Invicta Mirror Topic from Invicta Plastics

Invicta Tangram Work Cards from Invicta Plastics

Miras and *Mira Math Activities for Elementary School* from the Mira Math Co.

Pattern Blocks from Adelphi

Polyhedra Dice from Creative Publications

Pupils' Geo-Board Set and Geo-Board Activity Sheets from Ideal School Supply Co. ♥

Teacher Education

Using Microcomputers with Preservice Teachers

By **Thomas L. Schroeder**
University of Calgary, Calgary, AB T2N 1N4

A pressing issue facing teacher educators today is what work with microcomputers should be included in the program of preservice education for teachers. How we answer this question with respect to courses in mathematics education depends on our assessment of the skills, abilities, and attitudes we think future teachers need to develop and on such factors as the time and facilities available to us and the relationship of the mathematics course to the program as a whole—for example, whether the course is designed to deal with mathematics content or with the teaching of mathematics or with both. Although vast differences are bound to occur in the actions mathematics educators take in response to the needs and opportunities of their own situations, I believe whatever we do with microcomputers should be guided by the following three general considerations:

1. We should provide information and experiences that will broaden students' horizons by making them aware of the scope and diversity of potentially worthwhile uses for computers with elementary schoolchildren and their teachers.

2. We should ask questions that will help prospective teachers to become discriminating in selecting, designing, and implementing activities with computers.

3. We should encourage and model positive yet realistic and open-minded attitudes toward computers and their use.

Broaden students' horizons

Many people, especially those who have limited experience with computers, have serious misconceptions about educational computing. In our work with preservice teachers, we should be aware of these misconceptions and do what we can to help our students form balanced and realistic opinions.

One misconception is that if we buy the right computer hardware and software, then computers will take over a large part of the teaching now done by teachers. This view needs to be challenged. For one reason, it does not take into account the fact that teaching consists of a range of specific functions including diagnosing, explaining, giving examples, providing practice, questioning, and testing (to name just a few) and the fact that computers are probably much better suited to some of these functions than to others. Furthermore, the task of using judgment to coordinate these functions and decide when and how much to perform each one is a task that today's microcomputer systems cannot claim to do well. A second reason to challenge this view of computers is based on Papert's (1980) often-quoted distinction between situations in which "the computer programs the child" and ones in which "the child programs the computer." I believe that use of appropriate courseware should not be glibly written off, but neither should it be the only use (or even the major use) that preservice teachers observe.

Another misconception about educational computing is that in order to use a microcomputer, both teachers and students will have to learn a great deal about computer programming, particularly about BASIC. When I first began planning and conducting microcomputer workshops for teachers, I unquestioningly assumed that the major component would have to be instruction in BASIC. Since then I have deemphasized programming to allow more time for considering other educational uses of computers and to ensure that students who are slow catching on to the language do not come to the conclusion that using the computer is not for them. In most courses we cannot expect to teach everything we would like to teach about a language, but we can provide brief but positive experiences that will make students want to continue learning on their own.

A third misconception has to do with the use of games. When we mention microcomputer games, many people will think first of arcade games like Pac-Man and Space Invaders, and in this frame of mind some serious doubts and reservations about the educational value of games can be expected. Even if the games we advocate have more conspicuous mathematical content, a danger still exists that games will be valued mainly for the fun they offer. Teachers need to consider how they can capitalize on the mathematical experiences that gaming can offer. This approach means not only selecting games on the basis of the mathematical skills and concepts that are involved but also developing activities that will help children reflect on the mathematics they are learning and the problem solving that is involved in developing their own strategies for playing games.

Finally, some people view using a computer as an antisocial activity that can be hazardous to the user's physical and mental health. They fear that children will become hooked on technology and turn into glassy-eyed zombies who only want to communicate with a machine. Although granting that this image may fit a few extreme cases, let us also take note of the positive effects that often result when two or more people work together at a microcomputer. My experience with children and with adults has been that both the quantity and the quality of interaction is high as the partners ask each other questions, make hypotheses and suggestions, and discuss not only what happens but also why. Two studies, mentioned by Sweetham (1982), suggest that children interact more and cooperate better in computing activities than in regular classroom work. This finding may be due to a novelty effect, or it may be intrinsic to the computer, but in either case the result is beneficial and ought to be experienced and encouraged.

These and other misconceptions can be dispelled if we give our students a wide range of activities that mirror activities we believe are appropriate for children who are learning with computers. Many examples of such activities are discussed in the February 1983 issue of the *Arithmetic Teacher*, which focuses on teaching with microcomputers.

Ask questions

The activities we provide for our students will form the basis for their answers to the question, "How can a microcomputer be used in teaching?" But just because an activity is possible does not make it worthwhile. Other questions must be asked to help students think clearly about the nature, the suitability, and the potential for effectiveness of the different uses for microcomputers that they experience.

In examining available software, some relevant questions include the following:

- What are the purposes of the program? (e.g., to develop understanding? to practice a skill? to promote problem solving? to diagnose students' needs?)
- What is the content of the program? Can it be adjusted? How? By whom? When and for whom would it be appropriate?
- What features of the program did you like or dislike? (e.g., personalization; the content, amount, and timing of feedback; sound; graphics; repetition; variation and branching; record-keeping features)
- On what basis should programs for computer-assisted learning be evaluated and compared?

When we consider computer gaming, additional questions are needed. Many games can promote reflective thinking and problem solving on the part of the players, but often a well-chosen and well-timed question is needed to facilitate this effect. For example, questions like "Does it matter what move you make next?" and "What would be the best possible move you could make in this situation? Why?" can help players recognize the need for a strategy and help them initiate the process of developing, testing, and refining their strategies. These and similar questions are key elements of the activities that teachers should offer so that their students get the most benefit out of playing games. The questioning we do in teacher education courses should model the kinds of questioning we expect from our student teachers.

The teaching of programming can also be approached through questions; a major and recurring question is, "What commands can I give to the computer to get it to do what I want?" As future teachers consider the applications of this general question to specific classroom situations, other questions are relevant. What sorts of problems are suitable for children to solve by writing computer programs? What languages (e.g., BASIC, Logo, PILOT) and what commands in these languages should be introduced to children? When? In what sequence? When are the facts, concepts, and notations that children encounter in

learning computer languages similar to the facts, concepts, and notations of the mathematics curriculum, and when are they different? In what ways can we expect children's learning of computing to support their learning of mathematics, and vice versa? What can we do so that children's experiences in solving problems by computer, and especially their experiences in debugging programs, will contribute to their growth in problem solving in general?

Model positive attitudes

In the final analysis, how we teach may have as great an impact on our students as what we teach them. In view of the rapid rate of change in educational computing and the fact that our students may not begin their teaching careers for another year or two, an inquiring and open-minded attitude may be of more value to them than any particular information or specific prescriptions we might want to give. Both what and how we teach ought to reflect this reality.

References

Papert, Seymour. *Mindstorms: Children, Computers and Powerful Ideas*. New York: Basic Books, 1980.

Sweetham, G. "Computer Kids: The 21st Century Elite." *Science Digest* (November 1982):84–88. ◗

A Model for Preparing Teachers to Teach with the Microcomputer

By **Barbara R. Sadowski**

An Agenda for Action (NCTM 1980, p. 11) includes the following statement:

Teacher education programs for all levels of mathematics should include computer literacy, experience with computer programming, and the study of ways to make the most effective use of computers and calculators in instruction.

This recommendation provides a definition for the computer-literate teacher in the 1980s, a recommendation that would seem to be quite clear and straightforward except for a few not-so-obvious problems of semantics. Before deciding upon a program for preparing teachers, we need to clarify our position on these questions:

(*a*) What knowledge and skills define the computer-literacy competencies that should be required of specific types of teachers?

(*b*) What knowledge and skills should teachers have regarding the potential uses of microcomputers in the classroom?

(*c*) What should be our attitudinal goals for teachers, as the goals relate to microcomputers?

Computer-Literacy Goals for Teachers

Computer literacy means different competencies for different teachers,

Barbara Sadowski is director of research in the College of Education of the University of Houston. She teaches beginning and advanced instructional applications of the microcomputer, and conducts inservice teacher training in computer literacy. She also teaches graduate courses in diagnosis and remediation at the University of Houston Diagnostic Learning Center.

depending on the individual teacher's responsibilities. All teachers should know *about* the computer—its history, its impact on society, ethical questions involved in large computer data bases, and the nontechnical aspects of the limitations and capabilities of computers. Teachers also need to know how to integrate the computer into their teaching, which includes knowing the instructional uses of the computer and how to evaluate software. Evaluating software depends not only on knowing content and good teaching practices but also on understanding a programming language and how a program controls what the computer does.

Mathematics teachers at all levels, as well as teachers of gifted elementary children and computer specialists, should understand how a computer works and how to control it well enough to use it as a general-purpose tool for problem solving. These teachers should be able to read, understand, and modify programs; analyze problems; develop algorithms; flowchart (or develop a plan in psuedocode); and write and debug programs.

How Microcomputers Can Be Used

Teachers need a clear understanding of the various ways microcomputers can be used, so that they can make wise decisions on how to integrate computers into the curriculum. The microcomputer can be used to enhance instruction by (1) helping the teacher manage instruction (referred to as *Computer-Managed Instruction*), or (2) delivering the instruction

(*Computer-Assisted Instruction*), or (3) serving as the object of instruction (computer science and computer literacy).

Using the computer to manage instruction consists of keeping track of student progress toward objectives, diagnostic testing, recording wrong answers on CAI lessons, grading, recycling, and so on. Often a CAI package will include a student-management program so the categories of CAI and CMI are not always mutually exclusive.

The computer as a delivery system provides the most familiar and widest range of instructional uses. What the computer delivers, of course, is instruction by means of interactive programs called software that are stored on tape cassettes or diskettes. One author (Olds 1980) delineates three broad categories of software—tutorials, games, and simulations—but most computer educators would add drill-and-practice and problem solving that requires the user to write simple programs to generate data to investigate solutions. As such, problem solving is between CAI and computer science.

The computer is the object of instruction in both computer-literacy and computer-programming courses. Computer programming provides a unique opportunity to teach students how to use the computer as a general-purpose tool for problem solving. Higher-level cognitive skills of analysis, synthesis, and translation are called upon in the process of developing algorithms for the computer. Teaching students how to analyze a problem so that instructions can be written for the computer requires creative and well-thought-out teaching techniques. Knowledge about mathematics, a language for communicating with the microcomputer, and an understanding of how the computer interprets the program are necessary. An ability to reflect on one's own problem-solving strategy along with logical-thinking skill is more important than knowing a lot of mathematics or a high-level computer language. Knowing how to use this process of logically dissecting a problem-solving strategy, developing an algorithm, flowcharting, programming, and debugging is the key to an endless adventure with computers and learning.

Attitudinal Goals for Teachers

Teachers should see the computer as a versatile and powerful tool which aids both the teacher and the student in the mathematics classroom. It should be regarded as more than a device that provides drill and practice, averages grades, keeps track of student progress, or provides stimulating mathematics games, for it can be a tool to explore solutions to problems. The computer should not be viewed as just another electronic fad that will go away if teachers ignore it long enough. True, a computer is a machine that only does what it is programmed to do, but its interactive capability with response modification makes it unlike any other plug-in-the-wall teaching aid. Just as the computer is integrated into all facets of living, its use in the mathematics classroom should be as natural as using a pencil or calculator when appropriate.

Preservice Preparation of Teachers

There are two alternatives for preparing computer-literate teachers. A limited one is to address only the instructional uses of the computer as part of a methods course. A more satisfactory solution, given the broad goals of computer literacy, is a required course for all teachers, especially mathematics teachers and those planning to work with computers. An introduction to instructional uses of the computer, followed by a brief description of a computer-literacy course for teachers, follow.

Instructional uses of a computer

One way to introduce teachers to the instructional uses of the computer is to demonstrate software that illustrates CMI and each type of CAI.

Prior to the software demonstration, assign a reading that defines and compares the four types of software outlined earlier—tutorial, games, drill-and-practice, and simulations. (See Olds 1980, for example.) This reading assignment, which is then discussed in class, gives the student a framework or reference for the software demonstration that follows.

Computer vocabulary such as *hardware, software, disk, cursor, monitor,* and so on should first be defined. The software demonstration should be planned so that—

- the initial experience is as non-threatening as possible;
- each of the four types of software is demonstrated;
- several different content areas are covered;
- the full range of computer capabilities—graphics, sound, color, personalized random feedback, timing, user prompting, and a student-management section—should be included.

For example, the following set of programs might be used:

1. *Hello*. A friendly conversation that introduces users to the computer. (See Ahl 1978)

2. *Bagels*. A mathematics game that involves guessing a multidigit number. (Available from Minnesota Educational Computing Consortium)

3. *Metric Estimate*. A program that develops measurement concepts. (Available from MECC)

4. *Conics*. A program that draws conics from user specifications. (*Computing Teacher*)

5. *Speed Drill*. Timed drill on basic facts and more. (Available from MECC)

6. *Algebra*. A tutorial program on solving linear equations. (Available from MECC)

7. *Change*. A program for students that involves computing change and selecting coins. (Available from MECC)

8. *Sell Lemonade*. A simulation in which profits and losses are computed. (Available from MECC)

9. *Counting Bee*. A variety of programs, with a student-management file capability. (Available from EduWare, Inc.)

Before the software demonstration begins, several questions should be posed for the class to answer after each program is run.

1. What category of software is this?

2. What do you think the student learns from the program?

3. How and when would you use this in a classroom? Would you use it with individuals or with a group?

4. What features—graphics, sound, and so on—are used? Are they used appropriately?

5. What advantages are there for using the microcomputer to deliver this instruction? Could it be done as well or better by another method?

6. Are there any changes you would make to improve the effectiveness of the program?

To provide a basis for comparison, at least one software program that is viewed should have some questionable or poor-quality instructional practices—the same comment every time the response is correct, too many lines of print on the screen, no obvious way to exit the program, poor directions, inappropriate readability level, correct answers identified as wrong answers, distracting graphics, sound, or color, and so on.

The purpose of providing one or more poor-quality programs is to demonstrate that just because instruction is delivered via computers does not necessarily mean that it is high-quality instruction. Unfortunately, there are many examples of poor-quality instructional software available. Thus, the ability to critically evaluate software is one important skill that every teacher should develop, including the ability to critically examine the software in light of current instructional objectives. As an example, a program that provides *timed* drill on two-digit addition with regrouping is not consistent with current mathematics objectives. Metric estimation games, and practice and problem-solving strategy games, however, are consistent with goals for mathematics in the 1980s.

Because the user of a computer must know a computer language in order to instruct the machine, teaching problem solving to novices poses a real challenge. An example that works well is to present a program that simulates the tossing of one die 100 times, and printing out the total number of 1's, 2's, 3's, and so on. The teacher and students then examine the program listing to identify statements that (1) control the number of tosses, (2) "toss" the die, (3) store, and (4) print the numbers that come up. After each statement is explained, the discussion is focused on how one could modify the program to "toss" two dice 1000 times and print the sums of the numbers that come up. The modifications are made and results are compared to predictions.

A computer-literacy course would include more activities like the program modification just described, as well as other problem-solving examples. Enough programming should be taught to enable teachers to teach students to write simple programs for generating data to test hypotheses. In such a class, students would have the time and opportunity to learn more about computer literacy; instructional computing resources such as *Computing Teacher, Classroom Computer News,* and *Creative Computing;* in addition to simple problem analysis and programming, in-depth software

evaluation, and management of the computer classroom.

Inservice Computer Training

The following suggestions and observations are based on experience from over fifty inservice, hands-on, computer workshops conducted with teachers at every grade level.

1. Start with a brief, nonthreatening orientation, beginning with how to turn the microcomputer on, *unless you are sure that every teacher* has had this type of instruction. Definitions of basic computer vocabulary should be discussed also.

2. Have someone who is a computer novice sit at the keyboard and respond to a friendly program like *Hello.* An experienced teacher in this role only increases the anxiety of others.

3. After the introductory program, instruct the teachers on how to run a program from disk or cassette. Note differences between a typewriter keyboard and the microcomputer keyboard; and be sure to cover (a) how to respond when a question appears on the screen, remembering to press ENTER or RETURN, and (b) how to interrupt a program and run a different program. Suggest that they try to respond the way students might respond, just to see what happens.

4. Have groups of two to five teachers at each microcomputer to permit them to share their observations about the software. Let the teachers look at several types of software by exchanging disks or tapes if necessary. While the teachers run the software, answer any questions as you observe each group.

5. Reassemble the teachers and discuss any difficulties in running the programs. It helps alleviate anxiety when teachers know what some of the problems and error messages look like and what, if anything, they can do when these occur. No machine is 100 percent reliable, and the microcomputer is no exception. If some common error message was not encountered in running the software, then

describe the problem or error and its solution.

6. After the orientation, the direction of the next workshop or session should be set by the needs of the teachers. One option is to conduct a session like the one described for the preservice teachers, extending the discussion of types of CAI, and including an example of problem solving, such as finding the maximum area for a rectangle with a fixed perimeter.

7. If the teachers want to learn to program, there are at least two alternatives. For those who are not very comfortable with mathematics, dissect a program like *Conversation,* explaining the PRINT, INPUT, IF-THEN, and GOTO statements as they are encountered.

8. Another approach is to demonstrate several brief programs using the LET or assignment statement, GOTO, READ, PRINT, INPUT, and IF-THEN statements. Examples of such programs would be averaging a set of numbers, or generating the even numbers from 2 to 100, the squares and cubes of numbers from 1 to 50, and so on. Printing out the cubes of numbers using the exponentiation symbol (either $\uparrow$ or **) demonstrates a computer limitation that teachers need to be aware of. For example, $9 \uparrow 3 = 729.00000001$ because exponentiation is done with logarithms. Using INT $(9 \uparrow 3)$ will eliminate the problem.

Other workshops using the INT and RND functions to find prime numbers, simulate tossing dice, reducing fractions, and factoring numbers can follow.

A Final Word

As a teacher who has taught computer courses to students from the junior high school to the graduate college level, I have found that—

(1) *almost all students have been able to follow an example of the programming process, although not everyone has been able to carry out all of these steps independently.*

In the process of attempting to write a program to solve a problem, however, teachers gain extremely valuable insights into their own problem-solving strategies and their ability to instruct the computer. For in some ways a computer is the most literal of learners (exceptions are due to the hardware limitations) and as such, it does exactly what you tell it to do—not what you want it to do or what you think you have told it to do. At first, teachers find this literalness quite frustrating; but with help and encouraging support, most discover that they can do far more with the computer than they ever believed possible. This self-knowledge increases their confidence in themselves as teachers of mathematics.

(2) *the teachers who are able to program the computer are not always the most mathematically knowledgeable.*

Among the best programmers I have seen is a kindergarten physical education teacher, another a librarian; and one who now is a computer specialist in an elementary school began as a third-grade teacher. Not surprisingly, most mathematics teachers have very little trouble with learning to program the microcomputer.

(3) *contrary to the recommendations of computer-science professionals, it does not take several college-level mathematics courses and three or more computer-science courses for a teacher to teach children how to program a microcomputer.*

Teachers who have had a four-week intensive training in BASIC programming have gone on to teach elementary and secondary school classes in computer programming.

What seems to be necessary for this to happen is a willingness on the part of the teacher to relinquish the role of knowledgeable authority and to enter into a learning partnership with their students. As stated before, the cognitive skills are less important than the teacher's attitude toward computers and a sense of adventure about what can be learned by using the microcomputer. Not all teachers are willing or able to teach this way, but for those who are, learning and teaching about computers is likely to be the most exciting experience in their teaching careers.

References

National Council of Teachers of Mathematics. *An Agenda for Action: Recommendations for School Mathematics for the 1980s.* Reston, Va.: The Council, 1980.

Olds, H. "How to Think About Computers." In *Using Computers to Enhance Teaching and Improve Teachers Centers.* B. R. Sadowski and C. Lovett, eds. Houston: University of Houston, 1980.

Software sources

Ahl, D. H., ed. *Basic Computer Games (Microcomputer Edition).* Morristown, N.J.: Creative Computing, 1978.

Classroom Computer News, Box 266, Cambridge MA 02138.

Computing Teacher, c/o Computing Center, Eastern Oregon State College, LaGrande, OR 97850.

Creative Computing, P. O. Box 789-M, Morristown, NJ 07960.

Edu-Ware, Inc. P.O. Box 22222, Agoura, CA 91301.

Minnesota Educational Computing Consortium (MECC), 2520 Broadway Drive, St. Paul, MN 55113. ◗

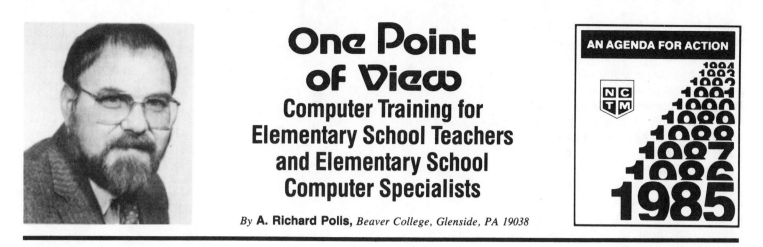

One Point of View

Computer Training for Elementary School Teachers and Elementary School Computer Specialists

By **A. Richard Polis,** *Beaver College, Glenside, PA 19038*

Many mathematics educators and specialists believe that computer education at the undergraduate level should be concentrated in the mathematics and mathematics/education courses. However, the content in these courses is already crowded, and the prospective elementary school teacher still needs more mathematics than is currently taught. To add the necessary knowledge of computers to this overcrowded curriculum is unwise for two reasons. First, it could dilute rather than enhance the mathematics training, and, second, computers should be applied to all disciplines. I believe that a general education in the use of the computer needs to be introduced first and that applications should then be integrated across the curriculum. This approach will allow and encourage teachers to use the computer as a tool in each of the elementary school subjects. This article recommends a program of computer education at the undergraduate level for elementary school teachers and elementary school computer specialists.

The elementary school teacher needs an outstanding education in the humanities and the social and natural sciences, as well as in the art and science of teaching. A broad liberal arts curriculum should provide the future elementary school teacher with

The Editorial Panel encourages readers to send their reactions to the author with copies to NCTM (1906 Association Drive, Reston, VA 22091) for consideration in "Readers' Dialogue." Please double-space all letters that are to be considered for publication.

the ability to think, make choices, evaluate, and solve problems and the motivation to continue as a lifelong learner. A portion of this education should include exposure to the technology of the present and the future. How much time should be devoted to computer technology given the brief four-year curriculum of thirty-five to forty-five college courses?

We can divide computer education into three levels: awareness, literacy, and competence-fluency. Clearly, every teacher—whether it be the art teacher or the social studies teacher—should have an awareness of what computers are and what they can do. At this level the prospective teacher needs to conquer the fear of the computer and should learn about the history of the computer as well as the present and future capabilities of hardware. The prospective teacher should have sufficient laboratory time to work directly on a microcomputer or should, on entrance to college, be encouraged to own a personal computer. Several colleges in the United States require students to own microcomputers and have purchased them at a price so low that almost every student can easily afford one. Requiring ownership makes a statement about the importance of the technology and allows for easy access without the frustration of waiting to use the college's computer.

Literacy requires that teachers be able to use and evaluate standard software, whereas competence-fluency requires that they be able to write and modify software. All teachers need to know what software is, how it can be used, and what criteria to apply in the

evaluation process. It would be ideal if all teachers could learn to modify and write their own software. In lieu of learning to program, a knowledge of PILOT would allow teachers to author their own software with little or no difficulty.

Every elementary school should have a well-designed curriculum that integrates the use of the computer in all subject areas, a reasonable amount of computer equipment, and at least one teacher who demonstrates competence-fluency in the field. This specialist should be called on as a resource person in each elementary school. Specialists should have a knowledge of computer languages, operating systems, and hardware and software, as well as the ability to design and modify software for teachers, help teachers to instruct children, and offer regular in-service training for teachers.

If all this knowledge can be integrated into the four-year curriculum for all students and particularly for prospective elementary school computer specialists, then teachers will immediately and naturally include the use of the computer as a tool in the elementary school. Finally, the way the teacher is taught to use the computer is crucial. This powerful tool can be used for solving problems better than any tool in our possession today, or it can be used for trivial exercises. The computer should not be used as a replacement for the workbook, rather, it should be used as a vehicle to help our children solve problems now and in the future.

Computer awareness can be compared to the knowledge of how an

SPECIFIC MATERIALS AND TECHNIQUES

163

automobile or a hammer is used and may include an understanding of how it works. One can easily understand the range of use and the capability of a tool without being able to use it. For many teachers this level may be sufficient. However, I believe that most teachers should be required to demonstrate a level described by many as "computer literacy."

The average person can drive a car and use a tape recorder, hammer, calculator, and so on. However, few people can repair a car or use the calculator to its utmost potential. To be literate one must have a keen awareness of the computer's capabilities, know when and how to use it, and be able to make judgments about the best uses for this tool. Literacy should include familiarity with a high-level language, as well as a knowledge of software and automated information systems. All teachers should know how to use the computer for word processing, filing, data search, and statistical analysis, as well as for problem solving. This knowledge might be accomplished in a course required of all college freshmen.

Level three, competence-fluency in the use of computer languages and operating systems, should be highly recommended for the elementary school teacher and be required of the elementary school specialist. Given the current state of the art and the languages used in schools, every teacher should be exposed to the languages of Logo, BASIC, and Pascal. At the same time, competence-fluency requires the ability to use several computer languages. I suggest Logo because of its successful use with young children and because of the ease with which it can be learned. BASIC is recommended because most current microcomputers seem to have BASIC built in. Pascal is an excellent choice because of its problem-solving capabilities, structure, and current use in public schools and colleges. The elementary school teacher needs to be familiar with Pascal, but the specialist needs to have a thorough knowledge of the language, including the ability to write structured programs and develop skills in problem solving. ◖

Teacher Education

Helping Preservice Teachers Develop an Awareness of Curricular Issues

By **Phares G. O'Daffer**
Illinois State University, Normal, IL 61761

A teacher refuses to allow calculators in the classroom because "the children won't learn their facts and computational skills." Another teacher elects not to read anything about computers, since "we don t have one in our classroom anyway." A third teacher skips the text's chapters on geometry, since "I don't see any real use for it." A fourth teacher "doesn't see why people think problem solving is so important" and calls it a passing fad. Another teacher wants a book that doesn't teach fractions because "we are in a metric age, and decimals are all that are needed." At the same time, another teacher wants a book that teaches only the customary units of measurement, since "the metric system didn't catch on."

It would be easy to discount these ideas by viewing these particular teachers as slightly out of focus on some of the current curricular issues. However, the main impact of their comments might be to cause us to ask the following:

As teacher educators, do we really help preservice teachers develop an awareness of the mathematics curriculum and a personal philosophy that will enable them to establish reasonable priorities when they are teachers?

During the past several years, one trend in methods courses for preservice teachers (PSTs) has been toward developing more practical ideas and techniques. Experiences such as planning bulletin boards, preparing activity cards, collecting games and puzzles, making manipulative teaching aids, watching videotapes about teaching, planning lessons, and increasing school participation are all extremely valuable. However, if we are to help PSTs develop an awareness of the curriculum and a working philosophy of mathematics instruction, we must offer them experiences that go beyond "what do I do on Monday?"

To increase their awareness of content and process goals, appropriate sequences, and major curricular issues, PSTs require a carefully developed set of experiences. Because of the limited time available in the methods class and because, at this level, PSTs tend to reject "theory," these experiences must be brief, vivid, and interesting.

The remainder of this article gives some specific learning experiences you may wish to provide for your PSTs.

Motivating textbook analyses

Since many beginning teachers rely on a textbook for curricular content and sequence, it seems reasonable to focus initially on this interpretation of the curriculum.

One technique that can motivate PSTs to examine textbooks and teacher's manuals carefully is to have them complete a "textbook scavenger hunt." Examples of some questions for the scavenger hunt are shown in figure 1.

Questions that are appropriate for the hunt should have the potential to heighten the PSTs' awareness of the major types of content taught at each grade level, developmental sequences, and the depth at which topics might be presented. Also, students can be encouraged to focus on the role of the textbook in handling important curricular issues related to basic computational skills, calculators, computers, problem solving, geometry, and the metric system.

After such an examination of a textbook series, one PST remarked, "I had no idea that so much is taught in third grade!" Other questions can focus on important things to be learned, such as concepts, facts, generalizations, skills, and the higher level processes of estimation, problem solving, and logical reasoning. The proportion of the program devoted to each type of learning can be assessed and noted for further discussion as the methods course progresses.

Note that the scavenger-hunt technique can also be used to analyze and compare curriculum guides in different districts.

Although this general look at the curriculum "through the eyes of a textbook series" is important, natural limitations in textbooks make it imperative that enlightened teachers use other sources to provide PSTs with a way to understand curricular issues and make intelligent decisions.

Making creative reading assignments

In methods courses, many mathematics educators have students read "how to" articles and prepare card files that contain practical ideas to be used when they begin teaching. However, a methods course can have so many things to do that students may have very little time for, or interest in, reading other articles that are more philosophical or theoretical.

Here are some ideas for brief "curriculum awareness" reading experiences that you may wish to adapt to your own setting:

1. Replace part of a regular exam with a take-home exam that includes an item like one of the following:

   a. Read pages 1–35 of John Holt's *How Children Fail.* Summarize the major pedagogical problems in the teaching and learning of mathematics that were presented. What are some of the things you could do as a teacher to minimize these problems?

   b. Read the introduction to *Mindstorms* by Seymour Papert. Give your reaction to the author's philosophy and major points. What is your philosophy regarding the use of computers in the elementary school classroom?

   c. Read Zalman Usiskin's "One Point of View" in the May 1983 issue of the *Arithmetic Teacher.* Do you agree with the author's position on the role of computational skills? Give reasons to support or rebut the arguments in the article.

   d. Read the chapter by Osborne and Kasten in NCTM's 1980 Yearbook, *Problem Solving in School Mathematics.* Rank the problem-solving goals and respond to the items on pages 54 and 55. Justify your rankings and compare them with the rankings of the groups in the study.

2. Create a letter that a parent might write on a current issue in mathematics education (inclusion of the metric system, hand-held calculators in the classroom, classroom use of microcomputers, etc.). Have your students prepare a thoughtful response to the letter, based on some study of the topic.

3. Select students to make ten-minute "Why Should We Teach It?" presentations on key topics suggested for inclusion in the curriculum by NCTM's *Agenda for Action,* the NCSM's *Basic Skills Report,* or other similar recommendations. Students

---

Fig. 1

Mathematics Textbook Scavenger Hunt

Name_____ Date_____

Select a textbook series (grades K–8) with a copyright date _____or later.

Name of series chosen _____

Publisher _____Copyright date_____

Authors_____

Study the texts and their teacher's editions to answer the following questions.

I. Some numeration concepts
   A. At what grade level are these concepts first explained?
      - three-digit place value   _____
      - expanded notation   _____
      - decimal place value   _____
      - comparison using > and <   _____
      - scientific notation   _____
   B. What types of models are used to show place value?
      _____

II. Computational skills
   A. At what grade level are these skills first introduced?
      - adding, with regrouping   _____
      - subtracting, with regrouping   _____
      - multiplying, three digits × one digit   _____
      - dividing with one-digit divisor, two-digit quotient   _____
   B. About what percentage of the_____grade book deals with computation?
      _____

III. Problem solving
   A. Are techniques or strategies for solving problems presented? Where?   _____
   B. Where does a problem *not* solved directly by +, −, ×, or ÷ appear?   _____
   C. Where do two-step problems first appear?   _____

IV. Calculators and computers
   A. Describe how calculators are used in the series. _____
      _____
   B. Does material on computer literacy appear? Where? _____

---

can find additional information for their presentations in the *Arithmetic Teacher* (e.g., in the February 1979 issue on the comprehensive curriculum), from other journals, or from such publications as *Selected Issues in Mathematics Education* (McCutchan Publishing Corp. 1980). Follow this activity with a class discussion of these issues.

Providing time for presentation and discussion of issues

Much has been said recently about on-task time and the importance of making every minute count in the elementary school. As teacher educators, we can practice what we preach about good use of class time. Why not use some of those minutes at the beginning or end of a class to discuss some current issues in mathematics education? For example, key points in a newspaper article about computers in schools can be highlighted. PSTs can be asked to give three or four words that come to mind as the points are presented. This brief word-association exercise might provide a springboard for a minidiscussion. Another newspaper article about a new program for gifted students in the local schools might be discussed. Interesting items from the "What's Going On . . ." section of the *Arithmetic Teacher* can provide an awareness of curricular developments. Even a remark about an important issue you've read about or a meeting you've attended is fair game for focusing on issues. An announcement about the theme and program emphases in upcoming local, regional, or national mathematics conventions would also be in order. This approach might go a long way toward instilling a desire to keep informed in PSTs. This desire, in turn, will contribute to the continued growth of teachers and an ever broadening awareness of the curriculum.

Another way to stimulate interest in major issues is to encourage PSTs to communicate with teaching colleagues. A mathematics convention is a great place for such communication! One methods instructor gives students credit for attending conference sessions and reporting on their experiences. Some instructors plan field trips to nearby conventions. A lot can be learned by PSTs about curricular issues by studying the topics presented at a mathematics education convention.

I would be remiss if I did not mention how history can provide a perspective on the ongoing process of curricular development. Many PSTs are surprised to learn that for decades a variety of groups (SMSG, UICSM, etc.) have wrestled with the *what, where, when, why*, and *how* of helping children learn mathematics. It is important for PSTs to understand, for example, that the current emphasis on problem solving is not new but a part of the cyclical pattern of concerns as the mathematics curriculum evolves.

Why not risk chuckles from academia and make an active-involvement bulletin board in your classroom to provide a brief encounter with the history of mathematics education for your PSTs?

Concluding remarks

It is a tall order to deal adequately with the needs of PSTs. Yet the effort and creativity we bring to this task may have a far-reaching impact. Such efforts may well mean the difference between a PST who continually improves as a teacher and who is excited about teaching mathematics and another PST who never has a significant commitment to the task.

Let's do our best to give PSTs the information and practical techniques they need for teaching mathematics, but let's also help them develop an awareness of the curriculum that is necessary in helping children learn mathematics.

References

Holt, John. *How Children Fail.* New York: Pitman Publishing Corp., 1964.

Lindquist, Mary Montgomery, ed. *Selected Issues in Mathematics Education.* Berkeley, Calif.: McCutchan Publishing Corp., 1980.

National Council of Supervisors of Mathematics. "Position Paper on Basic Skills." *Arithmetic Teacher* 25 (October 1977):19–22.

National Council of Teachers of Mathematics. *An Agenda for Action: Recommendations for School Mathematics of the 1980s.* Reston, Va.: The Council, 1980.

Osborne, Alan, and Margaret B. Kasten. "Opinions about Problem Solving in the Curriculum for the 1980s: A Report." In *Problem Solving in School Mathematics*, 1980 Yearbook of the National Council of Teachers of Mathematics, edited by Stephen Krulik and Robert E. Reys, pp. 51–60. Reston, Va.: The Council, 1980.

Papert, Seymour. *Mindstorms.* New York: Basic Books, 1980.

Usiskin, Zalman. "One Point of View: Arithmetic in a Calculator Age." *Arithmetic Teacher* 30 (May 1983):2. ◖

Teacher Education

Mathematics Educators:
Establishing Working Relationships with Schools

By **Paul R. Trafton**
National College of Education, Evanston, IL 60201

Mathematics education is a broad, diverse field. At the college level, mathematics educators engage in many professional activities, including instructing preservice and in-service teachers, conducting research, and addressing various problems and issues. As an applied discipline, mathematics education is directed toward the implementation of mathematically and educationally appropriate programs for students and the provision of high-quality mathematics instruction. If their efforts are to result in improved programs and instruction, mathematics educators need to have close, ongoing relationships with schools and with the day-by-day teaching of mathematics. Several reasons why these relationships are important can be advanced:

1. First-hand knowledge of the problems of teaching and learning mathematics in the actual classroom setting gives better focus to curricular discussions, identifies areas in which students need help, and provides insight into the concerns and problems of teachers. This knowledge is particularly important at the elementary school level, since few mathematics educators have extended, full-time teaching experience in the primary and intermediate grades.

A knowledge of the real world of the classroom also can increase the usefulness of preservice and in-service courses. Although these courses have many important objectives, they must provide useful, workable ways of dealing with existing problems. Suggestions for improving mathematics programs must also be perceived by teachers as feasible in their complex worlds.

2. Mathematics educators need to work regularly with students to test ideas and convictions, to find effective ways of helping students learn difficult topics, and to gain additional insight into children's thinking and learning processes. Access to students requires the approval and support of school administrators and teachers; such access requires that mathematics educators establish a working relationship with schools.

3. The preparation of preservice teachers in mathematics requires a variety of planned, focused clinical experiences. Many times clinical work is limited to student teaching, which, despite its importance, cannot accomplish all the goals of clinical work. Pre-service teachers need experience in working with students individually or in small groups. These clinical experiences provide manageable settings for learning specific teaching skills, for observing students' thinking processes, and for linking the content of methods courses to classroom instruction. Many times a level of control and focus can be built into clinical work that is not possible in student teaching.

Establishing a series of clinical experiences requires the cooperation of school administrators and classroom teachers, as well as their willingness to adjust the regular program to accommodate the college students. Experience indicates that it is feasible to develop the kind of working relationship with schools that makes clinical work possible and that these experiences have an important role in the preparation of mathematics teachers.

4. Mathematics educators have a responsibility to bring their knowledge and expertise to bear on the problems of mathematics programs in schools. Although these problems are not always of primary concern to the mathematics educator, the solution of them is a prerequisite to addressing broader questions about the mathematics education of students. Working in a local school is one way of reducing the serious gap between a knowledge of mathematics teaching and learning and the world of the classroom.

The need for mathematics educators at the college level to be involved with schools is clear, and school educators are frequently open to such involvement. Steps that mathematics educators can take to establish in-depth, continuing relationships with schools include the following:

- *Offering help to a local school or school system.* A meeting with a school principal about problems the school faces in teaching mathematics can lead to meetings with the whole faculty or a group of concerned teachers. Mathematics educators' willingness to work with schools on *their* problems is one way to build trust and confidence and open the door to more extensive involvement.

- *Working with a teacher on a regular basis.* Some mathematics educators arrange a working relationship with a single teacher; the mathematics educator becomes the mathematics teacher's coworker, both observing and participating. The arrangement has benefits for both individuals, because each learns from the other. From working and sharing together, a climate can develop that creates opportunities for testing new ideas and approaches. One variation of this approach is for the mathematics educator to teach the class for an extended period. The willingness of the mathematics educator to be closely involved with the school's program paves the way for other types of working relationships.

- *Volunteering to work with a small group of students.* An extra "pair of hands" is always needed in schools, especially with low- and high-achieving students. Although each group poses different problems, both cases involve concerns about teaching effectiveness, as well as appropriate content and materials. Thus, school personnel are usually open to the offer of outside help.

- *Involving teachers and schools in research projects.* Many research endeavors deal with students. Often, these projects require only that a few students be interviewed or taught by the research staff for a short period. Beyond this type of project, it is possible to design research efforts (perhaps as an extension of pilot work) that involve classroom teachers in testing materials and approaches on a larger scale. If school educators view the research as significant, they often are willing to participate. Again, both the mathematics educator and the teachers benefit, and teachers' interest and involvement in the project can lead naturally to other ways to cooperate.

- *Offering the assistance of preservice teachers.* Many states require that preservice teachers spend a large number of hours as classroom observers and participants before they start student teaching. The mathematics educator can arrange for a portion of this time to be spent in

selected schools. Having several students work in one school is a real benefit to the school and can be an initial step toward working with the school to design the type of clinical experiences described earlier. Regular visits by the mathematics educator when preservice teachers are in the school help establish communication and credibility with the staff.

Summary

Mathematics educators have a major stake in school mathematics. It is in the thousands of classrooms where the mathematics education of students is regularly carried out that change must be implemented if progress in mathematics teaching is to occur. Mathematics educators need to understand better the problems and constraints of the classroom. They need to be willing to work closely with schools on real-world problems and wrestle with ways of implementing change.

Schools also need outside expertise in addressing their problems. Too often, limited resources, limited personnel, and the need for quick solutions hinder teachers and schools in dealing with problems in a constructive, productive way.

Many ways exist for mathematics educators to establish relationships with local schools. Although working with schools places additional demands on the mathematics educator, the satisfactions and learning that result from such work make it a valuable, productive experience. ✎